Cosmic Explosions:
The Beasts and Their Lair

by

Edo Berger

ISBN: 1-58112-233-0

DISSERTATION.COM

Boca Raton, Florida
USA • 2004

Cosmic Explosions: The Beasts and Their Lair

Dissertation.com
Boca Raton, Florida
USA • 2004

ISBN: 1-58112-233-0

Cosmic Explosions: The Beasts and Their Lair

Thesis by
Edo Berger

In Partial Fulfillment of the Requirements
for the Degree of
Doctor of Philosophy

California Institute of Technology
Pasadena, California

2004

(Defended May 18, 2004)

Acknowledgements

Embarking on the fascinating study of gamma-ray bursts I could have found no better group of guides and companions than my past and present colleagues at the "Caltech-NRAO GRB group". I am indebted to them and many other colleagues at Caltech and elsewhere for their unconditional pursuit of scientific knowledge, and for making the last five years exciting and rewarding. I would like to think that some of this knowledge and excitement is captured in this thesis.

My thesis advisor, Professor Shri Kulkarni, has been an incredible and unrelenting source of knowledge, inspiration, motivation, emulation, on rare occasions frustration, scientific and political discussion (often charged and controversial), and friendly and professional advice. I thank Shri foremost for creating an environment in which dedication and hard work are both demanded and rewarded; for providing guidance, support and resources but at the same time thrusting (and trusting) me to find my way independently; and for his uncanny ability to dispassionately assess strengths and weaknesses, which has served to make me a better scientist and person. I really could not have wished for a better and more unique Ph.D. advisor, and I will always count Shri as a mentor, a colleague and a friend.

Although he has not been my official advisor, Dr. Dale Frail was involved in almost every aspect of my work from the very beginning. I thank Dale for an amazingly fruitful collaboration during the last five years , for introducing me to radio astronomy, for hosting me in Socorro during the summer of 2000, and for being a close colleague and friend despite the distance. I will forever remember the excitement and wonderful chemistry of our two-person radio afterglow team.

For their patience and friendship, for sharing their knowledge and expertise, and for bouncing ideas and schemes I thank the other members of the Caltech GRB group, Derek Fox, Alicia Soderberg, Josh Bloom, Paul Price, Sarah Yost, George Djorgovski, Fiona Harrison, Titus Galama, Dan Reichart, and recent arrivals Avishay Gal-Yam, Brad Cenko, and Dae-Sik Moon. I have learned much about GRBs, data reduction, and astronomy in general from all of them. I especially thank Re'em Sari for being a fountain of knowledge and a friend, always willing to share his wisdom; talking to Re'em has been like drinking from a firehose, but a few drops have been absorbed.

I thank the wonderful people at the VLA and AOC for providing and sustaining such an amazing instrument, and for supporting me as a young scientist when I made my first exciting discovery. I also acknowledge the great work and support of the staffs at the Keck and Palomar observatories, the James Clerk Maxwell Telescope and the Joint Astronomy Center in Hilo, and the Australia Telescope Compact Array and Australia Telescope National Facility. I have enjoyed visiting these facilities and interacting with the various people as much as I enjoyed the actual observing.

Many colleagues outside of Caltech have been instrumental in various aspects of this work. I thank the many collaborators and co-authors who donated their energy, expertise, and telescope time; their contributions are reflected and acknowledged in the individual chapters of this thesis. I thank the HETE, Beppo-SAX, IPN and INTEGRAL satellite teams for enabling the work I pursued in this thesis. Special thanks go to Len Cowie, who has been extremely generous in his support of my work on GRB host galaxies, and Ramesh Narayan who provided great insight into the workings of scintillation and has been a gracious host at the CfA.

At Caltech I have shared many discussions on GRBs, brown dwarfs, and the importance of statistics, as well as drinks and friendship with Bob Rutledge. Alice Shapley, Scott Chapman, Andrew Blain, Chuck Steidel, and Kurt Adelberger provided advice and invaluable information on high redshift galaxies, and Andrew MacFadyen has been a great resource for information on the physics and workings of collapsars. The members of my Candidacy and Thesis Defense committees, Fiona Harrison, Roger Blandford, Sterl Phinney, Andrew Lange, Sunil Golwala, Shri and Dale, provided great ideas for improving my thesis work.

While the last five years have been focused on research and astronomy, many friendships have made the process enjoyable and worthwhile. I thank the graduate students at the Caltech astronomy department for making life entertaining. In particular my classmates, Dave Kaplan and Micol Christopher, with whom I shared all the ups and downs of the last five years at Caltech. Dave has also been a constant resource of advice and insight into observing, data reduction, and MATLAB. I spent many lunches and post-Journal Club beers discussing politics, basketball, astronomy and life with Dave, Micol, Dawn Erb, George Becker, Bryan Jacoby and Ben Mazin. My officemates over the years deserve special recognition for putting up with me for endless hours day in and day out and for sharing their knowledge: Dave Vakil, Alice Shapley, Pat Udomprasert, Pranjal Trivedi, George Becker and Sean Moran. Over the years I have also enjoyed friendships outside of the Caltech astronomy community, Adi Adam, Roee Rubin, the Sari family, Elena Rossi, Davide Lazzati, the Dubowski family.

Of course, none of this would have happened without the love, support, and dedication of my amazing parents, Dorit and Arie, who continuously set an example and give me the tools for a happy, successful and fulfilling life; my wonderful brothers, Omri and Tom, whose warmth, intellect, sense of humor, and unbridled enthusiasm for their respective fields of study have provided a source of immense pleasure and happiness since the day they were born; and Alicia for being a truly unique person in every respect and for sharing her life with me.

Cosmic Explosions: The Beasts and Their Lair
by
Edo Berger

Abstract

The diversity of stellar death is revealed in the energy, velocity and geometry of the explosion debris ("ejecta"). Using multi-wavelength observations of gamma-ray burst (GRB) afterglows I show that GRBs, arising from the death of massive stars, are marked by relativistic, collimated ejecta ("jets") with a wide range of opening angles. I further show that the jet opening angles are strongly correlated with the isotropic-equivalent kinetic energies, such that the true relativistic energy of GRBs is nearly standard, with a value of few times 10^{51} erg. A geometry-independent analysis which relies on the simple non-relativistic dynamics of GRBs at late time confirms these inferences. Still, the energy in the highest velocity ejecta, which give rise to the prompt γ-ray emission, is highly variable. These results suggest that various cosmic explosions are powered by a common energy source, an "engine" (possibly an accreting stellar-mass black hole), with their diverse appearances determined solely by the variable high velocity output. On the other hand, using radio observations I show that local type Ibc core-collapse supernovae generally lack relativistic ejecta and are therefore not powered by engines. Instead, the highest velocity debris in these sources, typically with a velocity lower than $100,000$ km/sec, are produced in the (effectively) spherical ejection of the stellar envelope. The relative rates of engine- and collapse-powered explosions suggest that the former account for only a small fraction of the stellar death rate. Motivated by the connection of GRBs to massive stars, and by their ability to overcome the biases inherent in current galaxy surveys, I investigate the relation between GRB hosts and the underlying population of star-forming galaxies. Using the first radio and submillimeter observations of GRB hosts, I show that some are extreme starburst galaxies with the bursts directly associated with the regions of most intense star formation. I suggest, by comparison to other well-studied samples, that GRBs preferentially occur in sub-luminous, low mass galaxies, undergoing the early stages of a starburst process. If confirmed with future observations, this trend will place GRBs in the forefront of star formation and galaxy evolution studies.

Contents

1 **Introduction and Overview**   3
  1.1 History: The Discovery of Gamma-Ray Bursts and Their Afterglows 3
  1.2 Implications of a Cosmological Origin: Relativistic Fireballs 6
  1.3 Afterglows: Out of the Darkness and into the Light 7
    1.3.1 Radio Afterglows: Unique Diagnostics of the Energetics and Environments . . . 9
  1.4 Summary of the Thesis: Cosmic Explosions and Cosmology with GRBs 10
  1.5 Gamma-Ray Burst Energetics and Engine-Driven Supernovae 12
    1.5.1 What is the True Energy Release of GRBs? 13
    1.5.2 Are Local Core-Collapse Supernovae Driven by Engines? 14
  1.6 Cosmology with Gamma-Ray Bursts and Their Host Galaxies 16
    1.6.1 Are Dark Bursts the Key to Understanding Dust-Obscured Star Formation? . . . 16
    1.6.2 What Is the Nature of GRB Host Galaxies? 18

I **The Energetics of Cosmic Explosions**   21

2 **A Jet Model for the Afterglow Emission from GRB 000301C**   23
  2.1 Introduction 24
  2.2 Observations 24
  2.3 Data 25
  2.4 A Self-Consistent Jet Interpretation 27
  2.5 Conclusions 31

3 **GRB 000418: A Hidden Jet Revealed**   33
  3.1 Introduction 34
  3.2 Observations 34
    3.2.1 Optical Observations 34
    3.2.2 Radio Observations 35
  3.3 The Optical Light Curve and Host Galaxy 35
  3.4 The Radio Light Curves 35
  3.5 Global Model Fits 36

4 **A Standard Kinetic Energy Reservoir in Gamma-Ray Burst Afterglows**   43
  4.1 Introduction 43
  4.2 X-ray Data 45
  4.3 Beaming Corrections and Kinetic Energies 45
  4.4 Discussion and Conclusions 47

5 **The Non-Relativistic Evolution of GRBs 980703 and 970508: Beaming-Independent Calorimetry**   53

5.1	Introduction	53
5.2	The Non-Relativistic Blastwave and Fireball Calorimetry	55
5.3	GRB 980703	57
5.4	GRB 970508	58
5.5	Radiative Corrections	61
5.6	Discussion and Conclusions	61

6 A Common Origin for Cosmic Explosions Inferred from Calorimetry of GRB 030329 67

6.1	Radio Observations of GRB 030329	68
6.2	Broad-band Afterglow Models	69
6.3	A Two-Component Jet	70
6.4	A Common Origin for Cosmic Explosions	72

II   The Search for Engine-Driven Supernovae   75

7 The Radio Evolution of the Ordinary Type Ic SN 2002ap   77

7.1	Introduction	77
7.2	Observations	78
	7.2.1 The Radio Spectrum of SN 2002ap	78
	7.2.2 Robust Constraints	79
7.3	A Synchrotron Self-Absorption Model	80
	7.3.1 The SSA Model in the Context of a Hydrodynamic Model	81
	7.3.2 Interstellar Scattering & Scintillation	82
7.4	A Free-Free Absorption Model	82
7.5	Discussion and Conclusions	83

8 A Radio Survey of Type Ib and Ic Supernovae: Searching for Engine Driven Supernovae   85

8.1	Introduction	85
8.2	Observations	87
8.3	Population Statistics	88
	8.3.1 Radio Properties of Type Ib/c SNe	89
	8.3.2 Expansion Velocities	90
	8.3.3 Energetics	91
8.4	A Comparison to γ-Ray Burst Afterglows	92
8.5	Discussion and Conclusions	94
	8.5.1 What Is SN 1998bw?	96
	8.5.2 Hypernovae	97
8.6	Results for Individual Supernovae	98
	8.6.1 SN 2001B	98
	8.6.2 SN 2001ci	99
	8.6.3 SN 2002cj	99

III   The Multi-Wavelength Properties of Gamma-Ray Burst Host Galaxies   103

9 The Faint Optical Afterglow and Host Galaxy of GRB 020124: Implications for the Nature of Dark Gamma-Ray Bursts   105

9.1	Introduction	106
9.2	Observations	107

9.2.1 Ground-Based Observations 107
9.2.2 *Hubble Space Telescope* Observations 108
9.3 Modeling of the Optical Data 109
9.3.1 Cooling Break 109
9.3.2 Jet Break 110
9.4 Discussion and Conclusions 111

10 The Host Galaxy of GRB 980703 at Radio Wavelengths — A Nuclear Starburst in a ULIRG   **117**
10.1 Introduction 117
10.2 Radio Observations 118
10.3 Results 119
10.4 Evidence for Host Galaxy Emission in the Radio Regime 119
10.5 The Star Formation Rate in the Host Galaxy of GRB 980703 122
10.5.1 Star Formation Rate from the Radio Observations 122
10.5.2 Star Formation Rate from Optical and Submillimeter Data 123
10.6 Offsets and Source size 124
10.7 Discussion and Conclusions 124

11 A Submillimeter and Radio Survey of Gamma-Ray Burst Host Galaxies: A Glimpse into the Future of Star Formation Studies   **129**
11.1 Introduction 130
11.2 Observations 132
11.2.1 Target Selection 132
11.2.2 Submillimeter Observations 133
11.2.3 Radio Observations 134
11.2.4 Optical Data 134
11.3 Results 135
11.3.1 GRB 000418 135
11.3.2 GRB 980703 138
11.3.3 GRB 010222 138
11.3.4 GRB 000210 138
11.3.5 GRB 980329 139
11.3.6 GRB 000926 139
11.3.7 GRB 000301C 139
11.4 Spectral Energy Distributions 139
11.5 Star Formation Rates 141
11.6 Comparison to Optical Observations 142
11.7 Comparison of the Optical/NIR Colors of GRB hosts to Radio and Submillimeter Selected Galaxies 144
11.8 Conclusions and Future Prospects 146

12 Summary and Future Directions   **153**
12.1 The Diversity of Cosmic Explosions 153
12.2 Cosmology with Gamma-Ray Bursts and Their Host Galaxies 154
12.3 Conclusions 157

A Additional Publications   **161**
A.1 Refereed Publications Related to Gamma-Ray Bursts 161
A.2 Refereed Publications not Related to Gamma-Ray Bursts 164
A.3 Conference Proceedings 164

List of Figures

1.1 The fireball model of gamma-ray bursts 5
1.2 Radio to X-ray light curves of GRB 030329 exhibiting a two-component jet 7
1.3 Composite radio light curve demonstrating the physical insight provided by radio observations 9
1.4 A summary of our current understanding of GRB and type Ibc supernova energetics . . 12
1.5 Radio luminosities and expansion velocities of type Ibc supernovae 15
1.6 Optical light curves and upper limits from 110 searches in the period 1997–2003 17
1.7 Optical and near-infrared images of GRB 020127 18
1.8 $R - K$ color for GRB host galaxies as a function of redshift 19

2.1 Broad-band spectra of GRB 000301C 25
2.2 Radio lightcurves of GRB 000301C 26
2.3 Optical and near-IR lightcurves of GRB 000301C 29

3.1 Radio and optical light curves of GRB 000418 with ISM synchrotron models 37
3.2 Radio and optical light curves of GRB 000418 with Wind synchrotron models 38

4.1 Distributions of X-ray flux, luminosity and beaming-corrected luminosity for GRB afterglows 46
4.2 Isotropic-equivalent X-ray luminosity and γ-ray energy as a function of beaming factor . 48

5.1 Radio light curves of the afterglow of GRB 980703 with a fit to the non-relativistic evolution 56
5.2 Energies associated with the afterglow of GRB 980703 in the non-relativistic Sedov-Taylor phase as a function of the blastwave radius 58
5.3 Physical parameters of the Sedov-Taylor blastwave for GRB 980703 59
5.4 Radio light curves of the afterglow of GRB 970508 with a fit to the non-relativistic evolution 60
5.5 Energies associated with the afterglow of GRB 970508 in the non-relativistic Sedov-Taylor phase as a function of the blastwave radius 62
5.6 Physical parameters of the Sedov-Taylor blastwave for GRB 970508 63

6.1 Radio light curves of the afterglow of GRB 030329 68
6.2 Radio to X-ray light curves of the afterglow of GRB 030329 69
6.3 Histograms of various energies measured for GRBs 71

7.1 Radio light curves of SN 2002ap with synchrotron models 79
7.2 Explosion parameters for SN 2002ap 81

8.1 Radio light curves of Type Ib/c Supernovae 88
8.2 Radio light curves of Type Ib/c Supernovae and GRB afterglows 89
8.3 Blastwave expansion velocities of Type Ib/c Supernovae 91
8.4 Histograms of the radio luminosity of Type Ib/c Supernovae and GRB afterglows 93

8.5   Histograms of the radio luminosity of Type Ib/c Supernovae and GRB afterglows with
        models of the luminosity distribution   95
8.6   Histograms of γ-ray, X-ray and total relativistic energy for GRBs and Type Ib/c Supernovae   97

9.1   Palomar 200-inch and HST images of the field of GRB 020124   107
9.2   HST/STIS images of the faint optical afterglow of GRB 020124   108
9.3   Optical light curves of GRB 020124   110
9.4   Upper limits on optical emission from well-localized GRBs   111
9.5   Optical temporal decay index plotted against the R-band magnitude at $t = 1$ day   112

10.1   Radio emission from the host galaxy of GRB 980703   120
10.2   Variability of the radio flux from the host galaxy of GRB 980703   122
10.3   Angular offset of GRB 980703 relative to the radio host   125

11.1   Submillimeter and radio fluxes for GRB host galaxies as a function of redshift   136
11.2   Radio map of the host galaxy of GRB 000418 and its possible companion   137
11.3   Broad-band spectral energy distributions of several GRB host galaxies   140
11.4   Submillimeter/radio versus optical star formation rates for GRB host galaxies   143
11.5   Obscuration as a function of bolometric luminosity for GRB host galaxies detected in the
        submillimeter/radio   144
11.6 $R - K$ color as a function of redshift for GRB host galaxies   145
11.7   The projected sensitivities of EVLA, ALMA and $Spitzer$ shown with the SEDs of GRB
        host galaxies extrapolated to redshifts 1 and 3   148

12.1   Gamma-ray burst energies and afterglow X-ray luminosities plotted with the projected
        sensitivities of $Swift$ 154
12.2   Rest-frame B and K luminosities of GRB host galaxies   156
12.3   Hydrogen column density and MgII equivalent widths for QSO and GRB damped Lyα
        systems   157
12.4   The Broad-band SEDs of several GRB host galaxies along with the limiting sensitivities
        of future long-wavelength facilities   158

List of Tables

2.1   Radio and Submillimeter Observations of GRB 000301C   32

3.1   Optical/Near-IR Observations of GRB 000418   40
3.2   Radio Observations of GRB 000418   41
3.3   Synchrotron Model Parameters for GRB 000418   42

4.1   X-ray Afterglow Data   50
4.1   X-ray Afterglow Data   51
4.2   X-ray Afterglow Data at $t = 10$ hr   52

5.1   Physical Parameters of GRBs 980703 and 970508   65

6.1   Very Large Array Radio Observations of GRB 030329   73
6.2   Ryle Telescope Radio Observations of GRB 030329   74

7.1   Radio Observations of SN 2002ap   84

8.1   Radio Observations of Type Ib/c Supernovae in the Period 1999-2002   100
8.1   Radio Observations of Type Ib/c Supernovae in the Period 1999-2002   101
8.2   Ejecta and Progenitor Properties of Type Ib/c Supernovae Detected in the Radio   102
8.3   Best-Fit Models for the Supernova and γ-Ray Burst Luminosity Distributions   102

9.1   Ground-Based Optical Observations of GRB 020124   114
9.2   HST/STIS Observations of GRB 020124   114
9.3   VLA Radio Observations of GRB 020124   115
9.4   Afterglow Models for GRB 020124   115
9.5   Limits on Optical Afterglow Magnitudes for Bursts Localized in 2000–2002   116

10.1 Late-time radio observations of GRB 980703   127

11.1 Submillimeter Observations of GRB Host Galaxies   150
11.2 Radio Observations of GRB Host Galaxies   151
11.3 Star Formation Rates in GRB Host Galaxies Derived from Submillimeter and Radio
       Observations   152

To see a world in a grain of sand
And a heaven in a wild flower,
Hold infinity in the palm of your hand
And eternity in an hour

— William Blake (*Auguries of Innocence*)

What we call results are beginnings.

— Ralph Waldo Emerson

CHAPTER 1

Introduction and Overview

History: The Discovery of Gamma-Ray Bursts and Their Afterglows

Gamma-ray bursts are short, intense and non-thermal bursts of photons with $\sim$ MeV energies, which outshine the entire γ-ray sky (Klebesadel et al. 1973). GRBs were discovered serendipitously by the *Vela* satellites, launched in the late 1960s to monitor compliance with the Nuclear Test Ban Treaty, during a search for γ-ray emission from supernovae (Colgate 1968). The basic properties of these events were outlined within several years of their discovery (e.g., Cline et al. 1973; Wheaton et al. 1973; Strong et al. 1974; Imhof et al. 1974; Norris et al. 1984): (i) an apparently random distribution on the sky, (ii) durations ranging from less than 1 second to hundreds of seconds, (iii) a broken power-law energy spectrum with a maximum at a few hundred keV, (iv) a complex time structure resolvable on a timescale of at least several tens of milliseconds, (v) a relation between spectral hardness and intensity, and a softening of the spectrum as the burst evolves, and (vi) episodes of quiescence during which no emission above the background is detected. The GRB of April 27, 1972 (i.e., GRB 720427), detected on board *Apollo 16* (Metzger et al. 1974), provides an illustrative example: the burst lasted 25 s, exhibited pulse substructure on a timescale of 300 ms, a quiescent episode lasting several seconds, a possible precursor a few seconds prior to the main event, and a smooth power-law energy spectrum ranging from 2 keV to 5 MeV with a turnover at about 200 keV.

While most bursts share these basic properties, it is important to bear in mind that the GRB phenomenon is extremely diverse with durations, peak fluxes and fluences ranging over several order of magnitude, spectral peaks ranging from several keV (the so-called X-ray flashes or XRFs) to an MeV, and light curves ranging from highly variable to a smooth single peak. To date, only one simple classification scheme has been evident, with a class of long-duration ($t > 2$ sec), soft-spectrum GRBs, which account for about two-thirds of the known event rate, and a class of short-duration, hard-spectrum GRBs (Norris et al. 1984; Kouveliotou et al. 1993). This thesis is focused solely on the origin and diversity of the long bursts; the origin of the short bursts is perhaps the greatest unsolved mystery of GRB astronomy.

Following the discovery of GRBs, theoretical interpretations of the phenomenon spanned the gamut from stellar flares (Stecker & Frost 1973) to comets crashing into neutron stars (Harwit & Salpeter 1973) and "nuclear goblins" exploding upon ejection from their parent stars[1] (Zwicky 1974). The now-accepted association with the death of stars was first advanced by Bisnovatyi-Kogan et al. (1975) in the context of γ-ray emission produced by neutrino interactions during the stellar collapse process. By

---

[1] Curiously, Zwicky (1974) points out that nuclear goblins – putative parcels of nuclear matter stable only under extreme pressure (Zwicky 1958) – exploding with an energy of about 10^{40} erg would be accompanied by optical flashes of about 10th magnitude lasting about 100 seconds. Bright optical flashes were subsequently detected from a few GRBs (e.g., Akerlof et al. 1999), but the physical mechanisms and absolute luminosities of these flashes are entirely different than those envisioned by Zwicky.

1994 there were 135 published models for the origin of GRBs (Nemiroff 1994).

The proliferation of GRB models was a direct consequence of the unknown distance and energy scales. This resulted from the inability to collimate γ-rays and precisely associate GRBs with specific astronomical objects. Subsequently, source triangulation using several widely-separated spacecraft, the so-called inter-planetary network (IPN; e.g., Cline & Desai 1976; Barat et al. 1981), provided arcminute positional accuracy, but usually with a considerable time delay. An alternative, single-spacecraft approach was suggested by Gorenstein et al. (1976), using a wide-field, coded aperture mask hard X-ray instrument with a potential arcminute localization accuracy. The desired accuracy was driven by the need to localize GRBs to specific galaxies if they originated in the local universe, or specific bright stars if they originated in the Galaxy. Improvements in both designs ultimately provided the first localizations with sufficient accuracy and rapidity for the discovery of counterparts at other wavelengths.

Along with improvements in γ-ray positional accuracy, searches for counterparts in the radio and optical bands were also undertaken, as those could potentially provide arcsecond positions (e.g., O'Mongain & Weekes 1974; Grindlay et al. 1974; Baird et al. 1975, 1976; Cortiglioni et al. 1981; Schaefer 1986; Greiner et al. 1987; Hudec et al. 1987; Schaefer et al. 1989; Frail & Kulkarni 1995; McNamara et al. 1995). To some extent, these observations were motivated by theoretical predictions (see §1.3). However, based on our current knowledge it is clear that these searches did not reach sufficient depth rapidly enough.

The failure to implicate specific astronomical objects as the progenitors of GRBs turned attention to statistical methods, particularly the angular distribution of GRBs on the sky, and their number distribution as a function of peak flux, $\log N/\log S$. The former can vary between strong anisotropy, if GRBs originate within the disk of the Galaxy, to isotropy, if GRBs arise in an extended Galactic halo or are cosmological in origin (Usov & Chibisov 1975). The $\log N/\log S$ distribution follows the Euclidean power law slope $S^{-3/2}$ if the sources are uniformly distributed, but has a shallower slope at faint fluxes if the distribution is bounded in space (Prilutskii & Usov 1975). The alternative $V/V_{\max}$ test[2] (Schmidt et al. 1988) provides the same information, but it is not affected by the experimental sensitivity threshold.

The availability of degree-scale localizations, primarily from the IPN (e.g., Klebesadel et al. 1982; Hartmann & Epstein 1989), made it apparent that GRBs were not concentrated along the Galactic plane, and moreover were not associated with the Virgo cluster, nearby galaxies, or rich Abell clusters (van den Bergh 1983). The $\log N/\log S$ distribution was severely affected by the sensitivity threshold (Cline & Desai 1976), but the sample of bursts from the Venera 13-14 and Phobos missions did exhibit $\langle V/V_{\max} \rangle \approx$ 0.4, suggesting a deviation from uniformity (Mitrofanov et al. 1991). In a seminal paper, Paczynski (1986) used these preliminary results, along with the implied similar energy release to supernovae and an expected peak energy in the MeV range, to argue for a cosmological origin.

Significant progress, however, was made with the launch of the *Burst and Transient Source Experiment* (BATSE) on-board the *Compton Gamma-Ray Observatory*. The unprecedented sensitivity of BATSE combined with degree-scale localizations provided the first clear indication for a bounded distribution, $\log N/\log S \propto S^{-0.8}$ and $\langle V/V_{\max} \rangle \approx 0.35$, and an isotropic sky distribution (Meegan et al. 1992). These results were not consistent with known Galactic source populations and thus favored a cosmological origin (Paczynski 1991a,b). However, interest in Galactic models with an extended halo population ($d \gtrsim 20$ kpc; Li & Dermer 1992) remained strong, particularly in the context of the then newly discovered high-velocity neutron stars (e.g., Frail et al. 1994; Lyne & Lorimer 1994). The dispute between a Galactic and cosmological origin culminated in the "great debate" of 1995 (Lamb 1995; Paczynski 1995).

The long-awaited determination of the distance scale was finally made in 1997. On February 28

---

[2] Formally, $V_{\max}$ is the maximum volume to which an object can be detected in an experiment with a limiting count rate, $c_{\lim}$, and $V/V_{\max}$ is simply defined as $(c_{\mathrm{obj}}/c_{\lim})^{-3/2}$, where c_{obj} is the count rate of a particular object. For a uniform space distribution, $\langle V/V_{\max} \rangle = 0.5$.

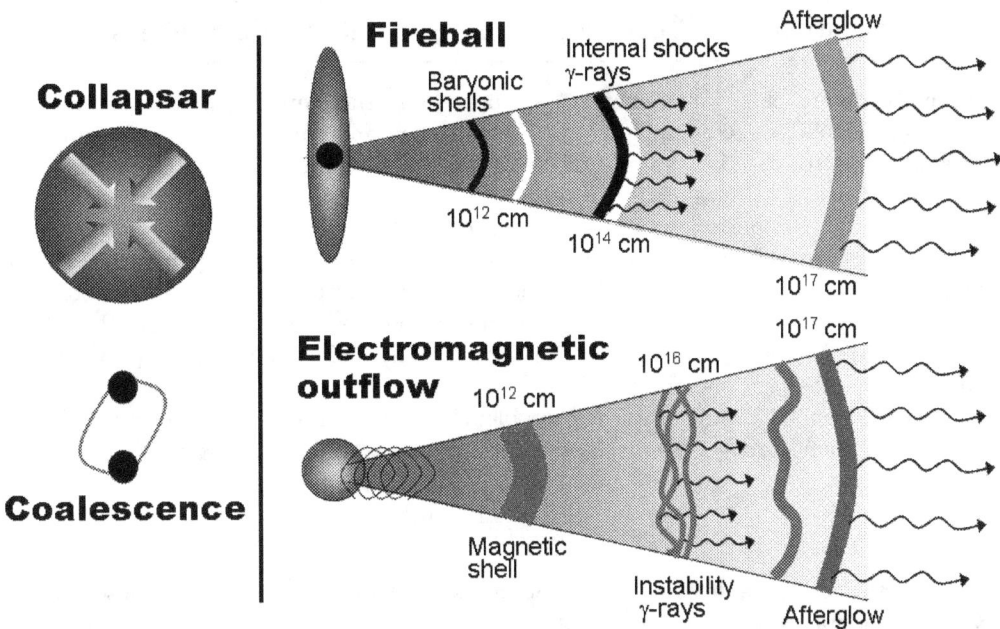

Figure 1.1: The energy scale of GRBs has focused attention on two progenitors models: coalescence of compact objects (NS-NS, BH-NS, BH-WD; e.g., Eichler et al. 1989) and "collapsars", accreting stellar mass black hole remnants, which power relativistic jets (Woosley 1993). Detections of supernova signatures in several long-duration GRBs supports the collapsar model, while coalescence events are thought to be the progenitors of the short-duration GRBs. Models for the energy transport focus primarily on fireballs (§1.2), in which the radiative energy is converted into kinetic energy of a low baryon load ($\sim 10^{-5}$ $M_\odot$) and is then re-converted to radiation via internal shocks (γ-ray burst) and an external shock with the circumburst medium (afterglow). Magnetic-dominated outflows are also possible, with a dissipation into γ-rays arising from magnetic instabilities.

of that year, the newly launched Dutch-Italian *Beppo-SAX* satellite (Boella et al. 1997) localized the prompt emission and fading X-ray afterglow (Costa et al. 1997) from GRB 970228 to a circle of 3-arcminute radius and relayed this information to ground observers a few hours after the burst. Optical observations revealed a fading afterglow (van Paradijs et al. 1997) associated with a faint source, shown with *Hubble Space Telescope* (HST) imaging to be extended and similar to a high redshift galaxy (Sahu et al. 1997). A direct confirmation of the cosmological origin was made with the next burst, GRB 970508, for which an absorption spectrum indicated a minimum distance of $z = 0.835$ (Metzger et al. 1997). By the time of writing of this thesis 35 GRB redshifts have been measured, ranging from 0.1 to 4.5, with a median redshift of $z \approx 1.1$.

One notable exception is GRB 980425 associated with the type Ic SN 1998bw at a distance of only 40 Mpc (Galama et al. 1998a; Pian et al. 2000). Due to its small distance, the γ-ray energy release of this burst was orders of magnitude below that of cosmological GRBs, while the associated SN 1998bw exhibited peculiarly high expansion velocities and kinetic energy compared to other type Ibc supernovae (Iwamoto et al. 1998; Kulkarni et al. 1998; Höflich et al. 1999; Li & Chevalier 1999; Nakamura et al. 2001). The origin of this GRB/SN is still hotly debated and has recently received great impetus with the detection of SN 2003dh, a close analogue to SN 1998bw, in association with the cosmological GRB 030329 (Hjorth et al. 2003; Stanek et al. 2003).

SECTION 1.2

Implications of a Cosmological Origin: Relativistic Fireballs

Gamma-ray bursts are one of the most energetic phenomena in the Universe, with isotropic-equivalent energy releases in some cases exceeding 10^{54} erg. Taken in conjunction with their short durations and high-energy spectra, GRBs require the violent creation of high energy photons in a compact region, a so-called fireball (Cavallo & Rees 1978). In general terms, fireballs are opaque due to the creation copious numbers of electron-positron pairs, and thus expand and cool until the energy spectrum is degraded below the pair-production threshold and the fireball becomes transparent. In the context of pure radiation fireballs, the fluid expands under its own pressure and so the bulk Lorentz factor, $\Gamma = [1 - (v/c)^2]^{-1/2}$, increases to relativistic velocities as the outflow becomes optically thin (Goodman 1986; Paczynski 1986). However, the emergent radiation is a thermalized blackbody spectrum, in direct contradiction with the observed spectra of GRBs.

This discrepancy, dubbed the "compactness problem," is at the heart of our current understanding of GRBs. The optical depth arising from the production of electron-positron pairs is

$$\tau_{\gamma\gamma} = \frac{f_{\mathrm{pp}}\sigma_T F d^2}{m_e c^2 R^2} \approx 10^{13}, \tag{1.1}$$

where $R = c\delta t \approx 3 \times 10^8$ cm is determined by the millisecond variability timescale observed in many GRBs, f_{pp} is the fraction of photons capable of creating pairs, and $F \sim 10^{-7}$ erg cm^{-2} and $d \sim 10^{28}$ cm are the fluence and distance of the burst, respectively. Clearly, the radiation will be thermalized.

The resolution of the compactness problem lies in the relativistic expansion of the fireball. For an outflow velocity Γ, the radiation is emitted from a radius $\Gamma^2 c\delta t$, while the photon energies in the rest-frame are lower by a factor of Γ. Thus, the optical depth is reduced by a factor of $\Gamma^{4+2\alpha}$ (where $\alpha \sim 1$ is the γ-ray spectral index), and for $\Gamma \gtrsim 100$ it is less than unity, giving rise to a non-thermal spectrum.

In the simplest scenario, and the one generally accepted at the present, the relativistic motion is intimately related to the dynamics of the fireball and the production of γ-ray radiation[3]. This was first understood in the context of the "baryon loading problem". Astrophysical fireballs are expected to entrain baryons, which will precipitate the conversion of the radiative energy into kinetic energy and will furthermore increase the optical depth of the fireball due to the associated electrons (Cavallo & Rees 1978; Goodman 1986; Shemi & Piran 1990). The relevant parameter in this context is the initial ratio of radiative energy to the rest mass energy of the entrained baryons,

$$\eta \equiv \frac{E_0}{Mc^2}. \tag{1.2}$$

The final Lorentz factor of the baryon-loaded outflow and the fraction of initial energy emitted as γ-rays both depend on the value of η, and are lower for increasingly large baryon loads (i.e., lower values of η). Thus, even for a load of $\sim 10^{-9}$ M$_\odot$ the delay in reaching optical thinness and the conversion of radiation to bulk motion result in a weak burst; for $M \sim 10^{-5}$ M$_\odot$, a GRB will not be produced at all, but the baryons will attain $\Gamma \approx \eta \gtrsim 100$.

To produce a GRB, therefore, the kinetic energy of the baryons has to be re-converted to radiation. This is achieved via deceleration of the ultra-relativistic outflow and dissipation of the kinetic energy in shocks, either externally by sweeping up interstellar matter (Meszaros & Rees 1992) or internally through instabilities in the outflow (Narayan et al. 1992; Rees & Meszaros 1994; Paczynski & Xu 1994); see Figure 1.1. A high value of Γ, and hence η, will give rise to γ-ray radiation. This naturally solves the compactness problem as well.

Thus, the unavoidable contamination of the fireball by baryons provides a mechanism for delaying

---

[3] Bulk relativistic motion of the source itself has also been considered (Krolik & Pier 1991), but this scenario is energetically unfavorable.

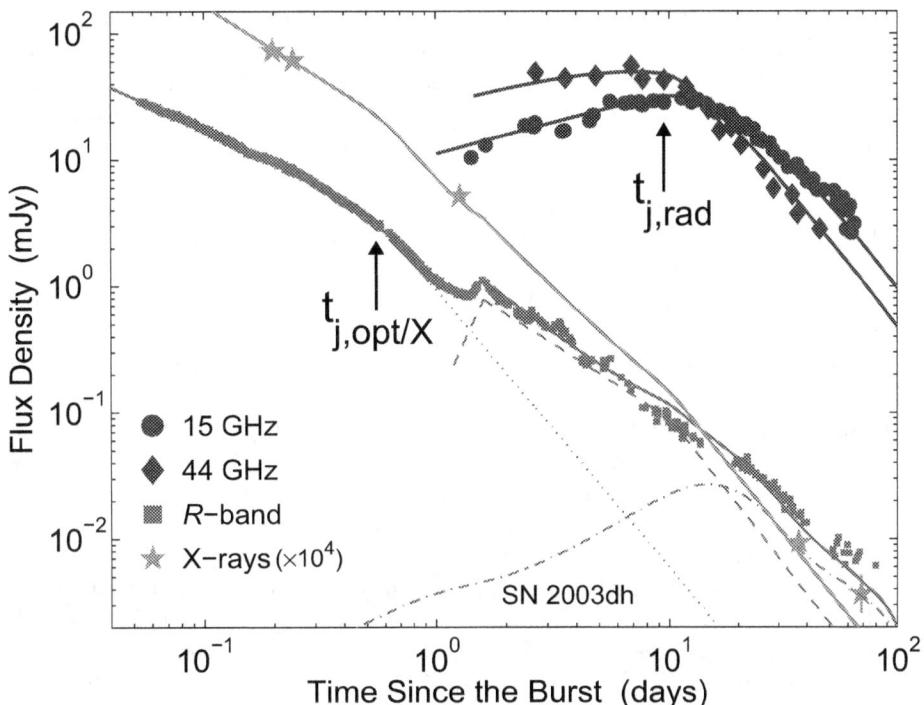

Figure 1.2: Radio to X-ray emission from the afterglow of GRB 030329 (Chapter 6). The solid lines represent models of synchrotron emission from a jet expanding in a medium of uniform density. The excellent match between the model and observations represents the success of the afterglow model in describing the observed properties and evolution of GRB afterglow.

the production of γ-rays until $\tau_{\gamma\gamma} \lesssim 1$ and at the same time provides a mechanism for the production of γ-rays. Efficiency arguments and the observed variability of GRB light curves implicate internal shocks rather than an external shock (but see e.g., Dermer & Mitman 1999). While the original baryon loading problem has actually provided a solution to the compactness problem, its current incarnation still persists, namely, how to ensure the right amount of baryons without producing a non-relativistic outflow with $\eta \sim 1$.

It is important to note that the outflow can alternatively be electro-magnetically dominated (e.g., Usov 1992); see Figure 1.1. In such models, the energy is extracted from the rotation of a strongly magnetized compact object or a black hole surrounded by an accretion disk. The relativistic magnetic outflow eventually develops instabilities which accelerate electrons and positrons and gives rise to γ-rays. Such models may have certain advantages over a baryonic fireball and internal shocks (Lyutikov & Blandford 2003), but at the present it is difficult to observationally discriminate between the two models.

┌─ SECTION 1.3 ───

Afterglows: Out of the Darkness and into the Light

The generic picture of GRB production invokes three steps: a compact "engine" which gives rise to a radiation fireball contaminated with a small fraction of baryons, the conversion of this energy to bulk motion of the baryons, and its re-conversion to non-thermal γ-rays. The identity of the engine is lost in the process, although some of its properties may be indirectly inferred from the prompt emission. Since the re-conversion of the kinetic energy to radiation is not fully efficient, a natural consequence of this scenario (and the one in which the outflow is magnetically-dominated) is the production of long-wavelength, long-lived emission as the fireball sweeps up and shocks the circumburst medium — an "afterglow" (Figure 1.1). Given the greater ease of observing X-ray, optical and radio emission and a significantly longer duration (days to weeks instead to seconds), observations of GRB afterglows have provided great insight into the properties of GRB engines and progenitors; an example is shown in Figure 1.2.

It is perhaps one of the most remarkable points about GRB research that the predictions of afterglow theory have held up so well when confronted with data. It is therefore worthwhile to outline the salient features of the production and evolution of GRB afterglows. Following the emission of γ-rays, significant deceleration of the relativistic shell, with Lorentz factor Γ_0 and an energy $E_K \sim (E_0 - E_\gamma)$, begins when it sweeps up $\sim 1/\Gamma_0$ of its rest mass energy. For typical values, $\Gamma_0 \sim 100$, $E_K \sim 10^{52}$ erg and a density $n \sim 1$ cm^{-3}, the deceleration begins about 90 seconds after the burst when the radius of the shell is about 10^{17} cm.

Assuming the shock does not radiate efficiently[4] the kinetic energy $E_K = 4\pi R^3 n m_p c^2 \Gamma^2/3$ is constant and thus $\Gamma \propto R^{-3/2}$. In addition, the observer receives emission from a given shell at $t \approx R/8\Gamma^2 c$ (Sari 1997), and therefore the evolution of the radius and Lorentz factor are given by $R \propto t^{1/4}$ and $\Gamma \propto t^{-3/8}$. This self-similar evolution was discovered by Blandford & McKee (1976).

We now have strong evidence that GRB outflows are collimated in jets (see §1.5). The exact hydrodynamic evolution of a jet with an opening angle θ_j has to be solved numerically, but in general terms it follows the spherical evolution outlined above so long as $\Gamma \gtrsim \theta_j^{-1}$. At later times, the outflow expands sideways under its own pressure, resulting in an exponential decrease of Γ as a function of radius. Thus, $R \sim$ const and $\Gamma \propto t^{-1/2}$ (Rhoads 1997, 1999; Sari et al. 1999).

The spectrum and evolution of the afterglow emission are determined by combining the dynamical solution with synchrotron radiation (e.g., Waxman 1997; Sari et al. 1998). The Lorentz factor of the post-shock fluid is $\Gamma_s = \sqrt{2}\Gamma$, the density is $4\Gamma_s n$ and the energy density is $4\Gamma_s^2 n m_p c^2$. The typical assumption is that the post-shock electrons are accelerated to a power-law distribution, $N(\gamma) \propto \gamma^{-p}$, above a cutoff $\gamma_{\min} \approx 300 \epsilon_e \Gamma_s$; the constant ϵ_e is the fraction of the shock energy that goes into the electrons. Similarly, a constant fraction is assumed to be contained in magnetic fields, $\epsilon_B = B^2/32\pi n m_p \Gamma_s^2 c^2$.

From basic synchrotron theory (e.g., Rybicki & Lightman 1979) and taking into account synchrotron cooling and self-absorption, integration over the electron distribution leads to a broken power-law spectrum; for example,

$$F_\nu \propto F_{\nu,m} \times \begin{cases} \nu^2 & \nu < \nu_a \\ \nu^{1/3} & \nu_a < \nu < \nu_m \\ \nu^{-(p-1)/2} & \nu_m < \nu < \nu_c \\ \nu^{-p/2} & \nu > \nu_c, \end{cases} \tag{1.3}$$

where ν_a is defined by the condition $\tau_{\mathrm{syn}}(\nu_a) = 1$, $\nu_m \equiv \nu(\gamma_{\min})$ is the frequency corresponding to the bulk of the electron population, $\nu_c \equiv \nu(\gamma_c)$ is the cooling frequency, and γ_c is the critical Lorentz factor above which the electrons lose a large fraction of their energy on the timescale of the system. The values of these break frequencies and the flux normalization are determined by the four basic parameters, E_K, n, ϵ_e and ϵ_B; other orderings of the break frequencies than those in Equation 1.3 result in different spectral shapes (e.g., Granot & Sari 2002). In the simple case of spherical expansion in a uniform medium $\nu_m \propto t^{-3/2}$, $\nu_c \propto t^{-1/2}$, $F_{\nu,m} =$ const and $\nu_a =$ const. Similar expressions have been obtained for a jet (Rhoads 1999; Sari et al. 1999) and for a blastwave expanding in a medium with a radial density profile (Chevalier & Li 1999).

The power of afterglow observations thus lies in the ability to directly infer the kinetic energy and

---

[4] During the first few hours of its evolution the blastwave is expected to lose about half of its energy to efficient synchrotron cooling (Sari et al. 1998). This will result in faster deceleration and slower expansion of the blastwave. Losses due to inverse Compton emission may further reduce the energy of the blastwave, but these are typically difficult to estimate given the paucity of X-ray data.

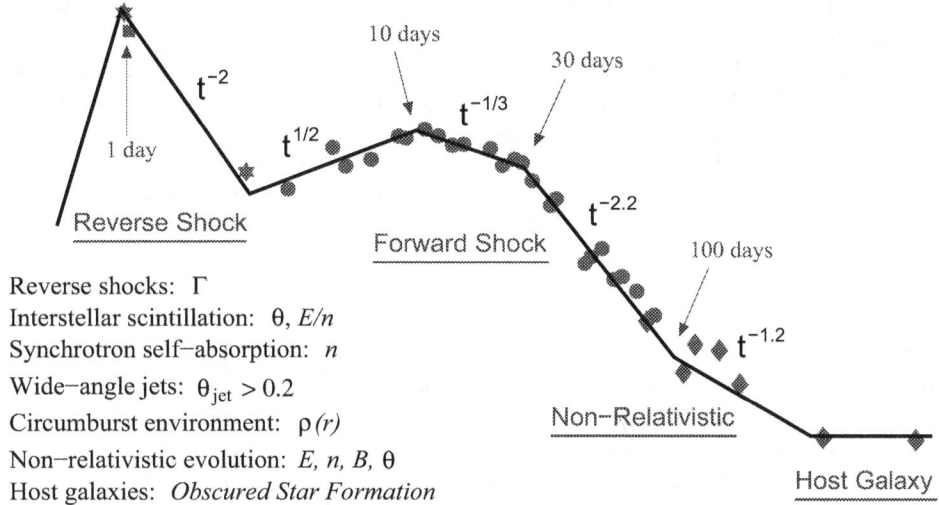

Figure 1.3: A composite afterglow light curve in the radio band scaled arbitrarily. Data are from GRBs 990123 (square; Kulkarni et al. 1999b), 020405 (stars; Berger et al. 2003d), 030329 (circles; Chapter 6) and 980703 (diamonds; Chapter 10). Timescales and scalings for the temporal evolution are indicated. The list summarizes aspects of the flux evolution which are unique to the radio bands and the physical insight they provide (Lorentz factor, Γ; source size, θ; energy, E; density, n; jet opening angle, θ_{jet}; density profile, $\rho(r)$; magnetic field strength, B; and obscured star formation rate).

geometry of the blastwave, the density and structure of the circumburst medium, and the micro-physical properties of the shock front. These parameters provide direct insight into the nature of the progenitor and the energy generation mechanism.

1.3.1 Radio Afterglows: Unique Diagnostics of the Energetics and Environments

Gamma-ray burst afterglows are a broad-band phenomenon requiring observations from radio to X-rays. However, the radio band provides some unique diagnostics of the afterglow physics and burst environment (Figure 1.3). To a large extent this is due to the slow evolution of the radio afterglow emission and its detectability for many weeks following the burst. This allows us to probe various phases of the dynamical evolution, as well as the burst environment over a factor of about ten in radius. Many of the results presented in this thesis take advantage of these unique aspects and I provide here a short summary (Figure 1.3).

At present, even with response times to GRB alerts of minutes, the radio band provides the best way to study the emission from the reverse shock (e.g., Soderberg & Ramirez-Ruiz 2003; Berger et al. 2003d), produced when the ejecta first decelerate (Sari et al. 1996). The properties of the reverse shock emission allow us to estimate the initial Lorentz factor. Optical observations require response on the timescale of the burst duration and have been made successfully only three times (Akerlof et al. 1999; Fox et al. 2003; Li et al. 2003). In the radio band, emission from the reverse shock has been observed several times since the peak happens about one day after the burst. In addition, the detection of reverse shock emission in the radio on this timescale most likely rules out a circumburst medium with a Wind (i.e., $\rho \propto r^{-2}$) density profile (Berger et al. 2003d).

The peak of the synchrotron emission from the forward shock is also missed in most optical and X-ray observations, and as a result these data alone cannot be used to infer the physical properties of the burst (Chapter 2). The radio band, however, directly traces the peak frequency and peak flux since those evolve through the band on a timescale of ~ 30 days. Moreover, only the radio band provides an estimate of the synchrotron self-absorption frequency, which is particularly sensitive to the density of the circumburst medium.

Radio observations are also well-suited for inferring the opening angles of wide jets, which are manifested at late time, $t \gtrsim 10$ days. On such timescales the host galaxy typically masks the optical afterglow, but in the radio band, the signature of such jets typically coincides with the peak synchrotron flux so wide jets are readily detected (Chapter 3). Similarly, since the long-term behavior is best studied in the radio, we can sometimes trace the transition to sub-relativistic expansion (Chapter 5), which occurs on a timescale of ~ 100 days (Frail et al. 2000c). These observations provide a beaming-independent estimate of the kinetic energy.

Finally, as the radio emission fades significantly we may detect emission from the host galaxy. At a typical redshift $z \sim 1$, the radio hosts detected to date have star formation rates in excess of 100 $M_\odot$ yr^{-1} (Chapters 10 and 11). These studies provide unique insight into the nature of GRB host galaxies and the environments most conducive for GRB progenitors.

Perhaps the most unique aspect of the radio emission is the existence of propagation effects in the form of interstellar scintillation (Goodman 1997; Walker 1998), which allow us to "resolve" the afterglow. These effects provided a confirmation of apparent superluminal motion (Frail et al. 1997), as predicted in the fireball model. More recently, very long baseline radio interferometry allowed us to resolve the afterglow of GRB 030329 and directly measure an apparent expansion velocity of $\sim 3 - 5c$ (Taylor et al. 2004).

SECTION 1.4

Summary of the Thesis: The Diversity of Cosmic Explosions and Cosmology with GRBs

Gamma-ray burst astronomy has matured considerably since the discovery of afterglows and the determination of the distance scale in 1997. Over the last few years we have addressed the preliminary question of what makes a GRB and we now know that they arise from the death of massive stars. Concurrently, in a manner reminiscent of quasar and type Ia supernova studies, I have focused on two paths: Understanding the diversity of these cataclysmic events, and using them as tools for cosmology. This thesis is thus motivated by two fundamental questions. First, *how diverse is the energy source driving cosmic explosions?* I address this question using afterglow observations to infer the true energy release of GRBs in the section entitled "The energetics of cosmic explosions". In the subsequent section, entitled "The Search for Engine-Driven Supernovae", I explore the relation between the two channels of stellar death using a radio survey of local type Ibc supernovae aimed at assessing their relativistic output.

The second question, *What is the relation between the host galaxies of GRBs and the population of field star-forming galaxies?* is addressed in part three, through a multi-wavelength study of GRB hosts and a comparison to other high redshift galaxy samples. This study provides additional insight, not available from optical studies alone, into the type of environments that may prove conducive to the formation of GRB progenitors.

Several methods are employed to attack the question of GRB energetics. In Chapters 2 and 3 I present the first use of full broad-band afterglow modeling to infer the physical properties of GRBs, along the lines of the discussion in §1.3. In both studies, the inference of jet collimation and the true energy release were only made thanks to the use of multi-wavelength data. A statistical study of beaming-corrected energies is discussed in Chapter 4, where I present a strong correlation between afterglow isotropic X-ray luminosities, a proxy for the fireball kinetic energy, and jet opening angles. The correlation indicates that for most bursts the kinetic energies are clustered within a factor of three. This clustering, coupled with a similar result from an analysis of the prompt γ-ray emission (Frail et al. 2001; Bloom et al. 2003b), places a quantitative constraint on central engine and energy extraction models. A beaming-independent assessment of the GRB energy scale using the radio emission from two bursts when the blastwave has decelerated to non-relativistic velocities is presented in Chapter 5. This analysis confirms that GRBs produce $\sim 5 \times 10^{51}$ erg, and independently validates the picture of strong collimation in GRBs.

Finally, I present in Chapter 6 detailed multi-wavelength observations of the nearby ($z = 0.168$) GRB 030329. The afterglow requires emission from two distinct collimated components. The first, which gave rise to the γ-ray and early afterglow emission, carried less than 10% of the total energy, $E_\gamma \sim 5 \times 10^{49}$ erg, while the second, mildly relativistic component dominated the late afterglow emission and had a typical energy. Following this example, we show that classical GRBs, XRFs and low E_γ events like GRBs 980425 and 030329 are unified through a common energy scale. This suggests that a single phenomenon is the culprit. The main difference between the various explosions appears to be the partition of energy between ultra-relativistic and mildly relativistic ejecta.

In a complementary effort to trace the diversity of stellar explosions, I present in Chapters 7 and 8 a comprehensive radio survey of local type Ibc supernovae designed to assess the fraction that are powered by an engine. This study was motivated by the association of GRB 980425 with the type Ic SN 1998bw. Two competing models for this event have been suggested: A typical GRB observed well away from the axis of the jet, and a new class of explosions perhaps straddling GRBs and typical core-collapse supernovae. I focus on type Ibc supernovae based both on the precedent set by SN 1998bw and on the understanding that their envelope-stripped progenitors can give rise to observable relativistic jets. I use radio observations because those provide a direct probe of relativistic ejecta. Based on the observations I reach four primary conclusions. First, the high-velocity output of type Ibc supernovae varies considerably, possibly reflecting a range of progenitor properties. Second, I place an upper limit of 3% on type Ibc supernovae that can be associated with GRBs or powered by engines based on the lack of detectable relativistic ejecta. Third, even if GRB 980425/SN 1998bw was a transition object, similar events comprise a small fraction of the total supernova rate. Finally, optical properties are poor indicators of an engine origin.

Part three of the thesis, entitled "The Multi-wavelength Properties of Gamma-Ray Burst Host Galaxies", presents the first radio and submillimeter observations and detections of GRB hosts, examines their properties in the context of other galaxy samples, and investigates the potential of GRBs for tracing dust-obscured star formation. In Chapter 9 I show that the lack of detected optical afterglows from the majority of the so-called dark GRBs is due to inadequate searches. As a result, the utility of GRBs in assessing the fraction of obscured star formation may be quite limited.

Chapter 10 revolutionizes our understanding of GRB hosts by extending their study to the radio band and by showing that GRB 980703 exploded within a nuclear starburst. Motivated thus, I have undertaken the first survey for radio and submillimeter emission from GRB host galaxies (Chapter 11). This study shows that several GRB hosts have star formation rates in excess of ~ 100 $M_\odot$ yr^{-1}, but these differ considerably from galaxies found in blank-field submillimeter surveys. In conjunction with optical and near-IR data I argue that GRBs likely arise in young starburst galaxies. This not only identifies GRBs as unique probes of recent cosmic star formation, but it also supports the consensus that GRBs arise from the most massive stars.

As is the case in all scientific endeavors, the studies described in this thesis raise many new questions on the diversity of stellar death and the nature of GRB host galaxies. The upcoming launch of the *Swift* satellite should provide new insight into these questions based on the increase in event rate and the more rapid and accurate localizations ($\sim 1-10$ arcsecond within a few minutes). This will naturally extend the potential of GRBs as unparalleled lighthouses for the study of the intergalactic medium and the interstellar medium of high redshift galaxies, and will sufficiently increase the sample of GRB hosts to allow a more meaningful comparison with other galaxy samples. We also expect that the higher sensitivity of *Swift* will uncover more low γ-ray energy events, and will settle the question of whether the observed clustering of the total energy is real or simply an artifact of sensitivity thresholds. I address some of these questions and future directions in Chapter 12.

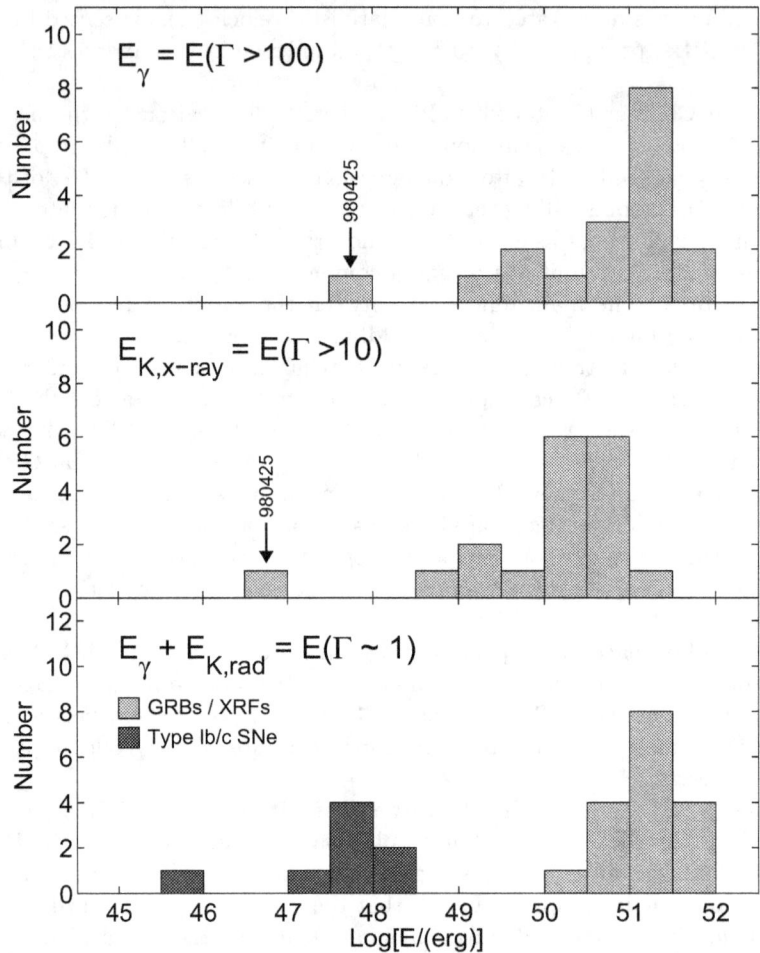

Figure 1.4: Histograms of various energies associated with cosmic explosions. *Top:* the beaming-corrected γ-ray energies tracing ejecta with $\Gamma \gtrsim 100$; *middle:* the beaming-corrected kinetic energy at 10 hours inferred from the X-ray afterglow ($\Gamma \gtrsim 10$); *bottom:* the total energy release including the kinetic energy inferred from the afterglow at late time. For the type Ibc supernovae this is the energy in the highest velocity ejecta ($v \sim 0.1 - 0.3c$) inferred from radio observations. The clustering of total energy is much stronger than E_γ or $E_{K,X}$ alone, indicating that in some cases the central engine channels the bulk of the energy in mildly relativistic ejecta. It remains to be seen whether the gap between cosmological GRBs and local type Ibc supernovae is occupied by intermediate energy explosions.

┌─ SECTION 1.5 ──
Gamma-Ray Burst Energetics and the Search for Engine-Driven Supernovae

The true energy release of gamma-ray burst engines depends critically on whether the outflow is spherical or narrowly-collimated. In the past, collimation was discussed both as a way of avoiding baryon loading (Ho et al. 1990; Krolik & Pier 1991; Meszaros & Rees 1992; Mochkovitch et al. 1993) and in the context of bulk relativistic motion channeled in outflows with an opening angle $\sim 1/\Gamma \sim 0.01$ rad. It was later recognized that the actual collimation of the jet can be larger than the angle defined by relativistic aberration, $\theta_j \gtrsim 1/\Gamma$ (Meszaros & Rees 1997; Rhoads 1997). Consequently, the prompt γ-ray emission does not allow us to assess the degree of collimation and the true energy release. However, as the outflow decelerates, the visible fraction of the jet surface grows larger, and when $\Gamma \sim \theta_j^{-1}$, a jet can be distinguished from a sphere. For $\theta_j \gtrsim 0.5°$, this happens in the regime of afterglow observations and is manifested as a break in the afterglow light curves.

This behavior has now been observed in several cases, with jet opening angles spanning from about 3 to 30 degrees (Frail et al. 2001; Chapters 2–6). Consequently, the true energy releases are potentially reduced by several orders of magnitude. This raises two crucial questions. First, what determines the opening angles of the jets, and are they correlated with other observables? Second, given that the true energy release is not so dissimilar from that of supernovae and that the GRB event rate is also increased by a factor of θ_j^{-2}, what is the relation between GRBs and supernovae?

1.5.1   What is the True Energy Release of GRBs?

The total relativistic energy produced by gamma-ray burst central engines is

$$E_0 = E_\gamma + E_{\mathrm{K,ad}} + E_{\mathrm{rad}}, \tag{1.4}$$

where E_{rad} is the energy radiated away in the early afterglow when a sizeable fraction of the electrons cool significantly, and $E_{\mathrm{K,ad}}$ is the adiabatic component which powers the long-lived afterglow emission. An unexpected result stemming from the inference of jet opening angles for several GRBs is that the distribution of beaming-corrected γ-ray energies is significantly narrower than that of the isotropic values: $E_\gamma \approx 1.2 \times 10^{51}$ erg, with a 1σ spread of about a factor of two (Frail et al. 2001; Bloom et al. 2003b); see Figure 1.4.

In principle both $E_{\mathrm{K,ad}}$ and E_{rad} can be measured from detailed afterglow data. In practice, most bursts do not have adequate coverage to fully constrain the energy (Panaitescu & Kumar 2002; Yost et al. 2003). A more robust approach to estimating the *distribution* of kinetic energies is available, using the early X-ray luminosity (Chapter 4). This method is based on the fact that the flux at frequencies above the cooling frequency (i.e., X-rays) is proportional to $\epsilon_e E_K$ (Kumar 2000; Freedman & Waxman 2001), but is independent of the circumburst density and depends very weakly on ϵ_B. Therefore, the distribution of beaming-corrected X-ray luminosities is a direct proxy for the distribution of true kinetic energies. For a sample of twenty GRBs with known jet opening angles, the isotropic X-ray luminosity is strongly correlated with the beaming fraction, such that the true X-ray luminosities, and hence $E_{\mathrm{K,ad}}$, are nearly constant (Figure 1.4). Thus, the wide dispersion in both $E_{\gamma,\mathrm{iso}}$ and $E_{\mathrm{K,iso}}$ is simply a manifestation of the diverse opening angles[5].

The reduced dispersion in true X-ray luminosity has other significant ramifications. Namely, since $\epsilon_e E_K \propto L_X Y^\epsilon$, with Y proportional to the isotropic X-ray luminosity and $\epsilon \equiv (p-2)/(p-1)$, the factor Y^ϵ should be nearly constant. Otherwise, there is no reason why L_X should be nearly constant. Thus, several conditions are necessary. First, the X-ray luminosity is dominated by synchrotron rather than inverse Compton emission since the latter depends sensitively on the density and ϵ_B, which vary considerably from burst to burst. Second, p must be relatively constant and have a value close to 2 to ensure that Y^ϵ does not vary significantly. Third, given that the combination $\epsilon_e E_K \approx$ const, this requires ϵ_e and E_K individually to be nearly constant. This would not be required if the two quantities are correlated, but there is no reason to assume that the shock microphysics depends sensitively on the kinetic energy. Finally, since both the prompt and afterglow emission are strongly correlated with θ_j, which is determined from afterglow observations, the standard energy result indicates that GRB jets are relatively homogeneous and maintain a simple geometry all the way from internal shocks ($\sim 10^{14}$ cm) to a radius of about 10^{17} cm. This analysis thus provides powerful constraints on the energetics,

---

[5] An alternative suggestion (Rossi et al. 2002) is that the inferred angles actually reflect the observer line-of-sight relative to the jet. In this case, GRB jets still have a standard energy, but they are structured with $E_\theta \sim \theta^{-2}$ and have the same opening angle in all cases. At present we are unable to distinguish between the two interpretations since the afterglow flux evolution in both models is nearly indistinguishable.

geometry and shock microphysics of GRBs.

While $E_{\rm rad}$ is difficult to estimate, the fact that both E_γ and $E_{\rm K,ad}$ appear to be nearly constant, indicates that $E_{\rm rad}$ is similarly distributed and probably does not represent a major fraction of the total energy budget. It therefore appears that $E_{\rm rel} \sim$ const with a value of few $\times\, 10^{51}$ erg.

Given the implications of the standard energy result, we would like to assess the energy content of GRBs *independent of assumptions about jet collimation*. Fortunately, the late-time radio emission affords such a tool since on a timescale $t_{\rm NR} \sim 65(E_{\rm iso,52}/n_0)^{1/3}$ d the blastwave becomes non-relativistic and approaches spherical symmetry even if it was initially collimated (Livio & Waxman 2000). Thus, I use the Sedov-Taylor self-similar solution to model the late radio emission from GRBs 970508 and 980703 and estimate the total kinetic energy of the fireball (Chapter 5). This approach has the added advantage that, unlike the γ-ray and X-ray studies, it can also trace any non-relativistic ejecta produced by the central engine. I find that $E_K \approx 5 \times 10^{51}$ erg, thus confirming the energy scale and jet collimation.

Alongside the standard energy yield, the γ-ray and X-ray analyses also highlight a group of sub-energetic bursts, including GRB 980425, whose energies in ejecta with $\Gamma \sim 100$ (γ-rays) and $\Gamma \sim 10$ (X-rays) are at least on order of magnitude lower than typical values (Figure 1.4). The relation between these bursts and the "classical" cosmological bursts was recently revealed through observations of the nearby GRB 030329 (Chapter 6). Detailed, high precision, observations of this burst in the radio, millimeter, submillimeter, near-IR, optical and X-ray bands have pointed to a two-component jet in which the bulk of the energy is in mildly relativistic ejecta (Figure 1.2). Thus, the central engine of GRB 030329 produced the standard energy yield, but a fraction of only 5% was channeled in ultra-relativistic ejecta. A close examination of other sub-energetic GRBs reveals a similar picture. In particular, in GRB 980425 the total relativistic energy yield was $\sim 10^{50}$ erg (Li & Chevalier 1999), with a fraction of only 1% in ultra-relativistic ejecta.

As summarized in the closing chapter of part I, the emerging picture is the following. Cosmic explosions (GRBs, XRFs and SN 1998bw-like events) appear to have a nearly standard energy yield with a about factor of three spread. However, the partition of energy between ultra-relativistic and mildly relativistic ejecta varies considerably, such that E_γ is a poor indicator of the total energy yield. This forces us to both revise our view of gamma-ray bursts as events which are energetically dominated by γ-rays, and also address the question: what physical parameter(s) related to the engine and/or progenitor control the partition of energy between various levels of baryon loading?

1.5.2   Are Local Core-Collapse Supernovae Driven by Engines?

The search for astronomical γ-ray transients in the *Vela* data was prompted by the suggestion that supernovae might emit a pulse of γ-rays when the shock front first breaks out of the exploding star (Colgate 1968). However, no γ-ray emission was detected in coincidence with any known supernova. Talbot (1976) points that the if the GRB and supernova rates are similar, then the lack of association indicates that either: (i) GRBs, rather then supernovae, dominate the stellar death rate, or (ii) all GRBs are associated with supernovae but those are too faint to detect since the distance scale is larger than about 100 Mpc.

The cosmological origin of GRBs rules out case (i); in fact the GRB rate, even with the most optimistic correction for beaming ($f_b^{-1} \approx 500$) is about 0.5% of the rate of type Ibc supernovae (Chapter 8). On the other hand, over the past several years photometric and spectroscopic signatures of supernovae have been detected in association with several cosmological GRBs (e.g., Bloom et al. 1999; Stanek et al. 2003). It remains to be seen whether *all* GRBs are associated with supernovae.

At the same time, the growing recognition that GRBs have a standard energy yield, has given rise to a spate of "unification models". The most extreme of these models (Lamb et al. 2004) posits that GRBs and XRFs have a common energy scale determined by the lowest energy event detected in the sample, XRF 020903 with a prompt energy release of only 10^{49} erg. The jet opening angles required to lower the energy scale to such values are a factor of about ten times smaller than the generally-accepted values. As a result, the GRB event rate is a factor of 100 higher than previous estimates, leading to the

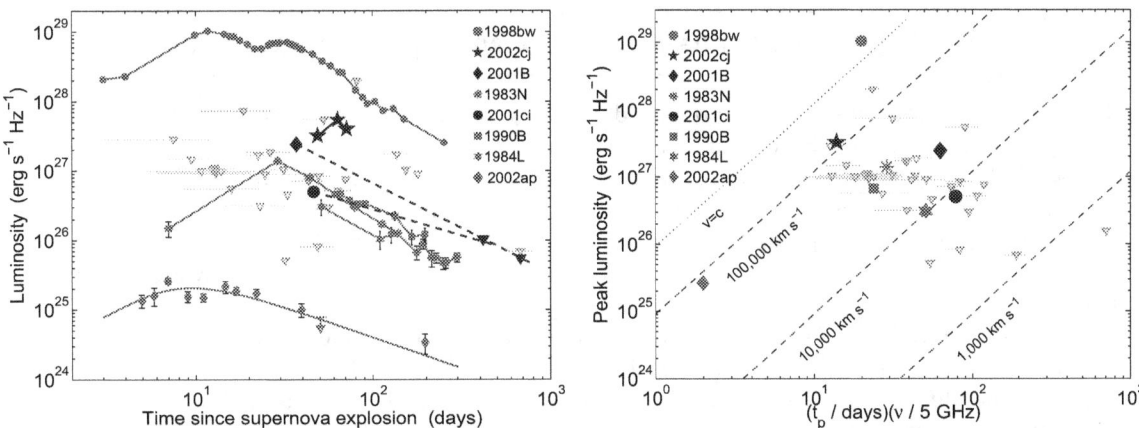

Figure 1.5: *Left:* Radio light curves of type Ibc supernovae and upper limits for the non-detections (triangles). The uncertainty in time for the non-detections represents the uncertain time of explosion. *Right:* Peak luminosity plotted against time of the peak for the same supernovae. The diagonal lines are contours of constant average expansion velocity (Chevalier 1998). Clearly, none of the sources observed to date were as luminous as SN 1998bw or exhibited relativistic expansion, suggesting that most type Ibc supernovae are not powered by engines.

condition that nearly all type Ibc supernovae give rise to GRBs — a true unification scheme.

We can assess such claims and shed light on the relation between GRBs and supernovae through studies of local type Ibc supernovae. If some of these supernovae are simply GRBs pointed away from us, then the fraction with strong radio emission (produced when the outflow is nearly spherical) is tied to the beaming angles and ranges from 0.5% to ~ 100%. However, an intermediate population of sources will be independent of the GRB rate, and one suggestion (Norris 2002) is that nearly 25% of all type Ibc supernovae should exhibit engine signatures.

The origin of SN 1998bw and similar events may therefore be assessed with radio observations of a large sample of type Ibc supernovae. To this end I conducted a radio survey of such supernovae between late 1999 and the end of 2002. This study shows that less than 3% of type Ibc supernovae are powered by engines (Figure 1.5). In fact, the high-velocity ejecta detected in some cases and the inferred energies (Figure 1.4) can be easily explained as the tail of the ejecta velocity distribution (Chapters 7 and 8). The wide range of radio luminosities, spanning at least four orders of magnitude, presumably reflects the sensitivity of high-velocity ejecta to the properties of the progenitor (e.g., size, density gradient). In addition, several supernovae which were classified as "hypernovae" based on their similarity to SN 1998bw in the optical band, lack strong radio emission. This indicates that the optical emission is not a reliable probe of an engine origin (Chapter 7). This may not be surprising given that the optical emission arises from radioactive decay of ^{56}Ni, whose production may not be unique to the explosion mechanism.

It is therefore apparent that only a minor fraction of local type Ibc supernovae are powered by engines, ruling out the claimed fractions of 25% (Norris 2002) and ~ 100% (Lamb et al. 2004). We still cannot distinguish models in which SN 1998bw-like events are off-axis GRBs with typical jet opening angles from those in which they are transition objects. However, even if the latter proves to be the case, such explosions represent only a small fraction of the local stellar death rate.

┌─ SECTION 1.6 ──┐
│ # Cosmology with Gamma-Ray Bursts and Their Host Galaxies   │
└───┘

The localization of GRB afterglows to arcsecond accuracy has enabled the detection of host galaxies underlying the burst positions. Studies of these galaxies have been focused on two primary paths. First, their detailed astrophysical properties provide indirect clues to the nature of GRB progenitors. Thus, observations indicate that GRB hosts are star-forming galaxies and may have relatively low metallicity, perhaps an indication that the production of GRBs favors such environments. In addition, the angular offsets of GRBs relative to the distribution of starlight, has been used statistically to favor massive stars as the progenitors (Bloom et al. 2002a).

Equally important, GRB host galaxies can be used to study the evolution of star formation and galaxy formation. In this context, present studies are still limited by the biases and shortcomings of optical/UV, submillimeter and radio selection techniques. In particular, optical/UV surveys may miss the most dusty, and vigorously star-forming galaxies, and it is not clear if the simple prescriptions for correcting the observed star formation rates for dust extinction (e.g., Meurer et al. 1999) actually work at high redshift. Submillimeter surveys have uncovered a population of highly extincted galaxies with star formation rates of several hundred $M_\odot$ yr^{-1} (e.g., Smail et al. 1997), but uncertain positions have made it difficult to measure their redshifts. Finally, studies in both the radio and X-ray bands suffer from contamination by active galactic nuclei. Perhaps the most severe limitation of all studies, particularly in the submillimeter and radio, is that they are flux limited and may potentially miss the bulk of the star formation if it occurs in faint galaxies.

Against this backdrop, GRBs afford a unique way of selecting high redshift galaxies in a way that overcomes some of these selection effects. In particular:

- The galaxies are selected with no regard to their emission properties in any wavelength band

- The dust-penetrating power of the γ-ray emission results in a sample that is completely unbiased with respect to the global dust properties of the hosts

- The redshift of the galaxy can be determined via absorption spectroscopy of the optical afterglow allowing a redshift measurement of arbitrarily faint galaxies (the current record-holder is the host of GRB 990510 with $R = 28.5$ mag and $z = 1.619$; Vreeswijk et al. 2001b)

- GRBs are detectable to very high redshifts, should they exist there ($z \gtrsim 10$; Lamb & Reichart 2000)

Naturally, GRB selection may have its own biases, but it is safe to conclude that GRB hosts provide a new perspective on star formation studies, which is at least subject to a different set of systematic problems than the optical/UV and submillimeter approach.

In addition to the insight afforded by host galaxy studies, it has also been suggested that the fraction of GRB afterglows strongly obscured by dust can act as a surrogate for the fraction of obscured star formation. This is of great interest since galaxy surveys in various bands, give rise to different conclusions (e.g., Madau et al. 1996), partly because they are based on secondary indicators, such as the amount of re-processed starlight and the amount of UV absorption. On the other hand, *so long as GRBs are not biased with respect to the cosmic star formation,* the GRB approach is direct and possibly offers the best estimate of obscured star formation.

1.6.1   Are Dark Bursts the Key to Understanding Dust-Obscured Star Formation?

One of the main observational results stemming from several years of GRB follow-ups at optical wavelengths is that about 60% of well-localized GRBs lack a detected optical afterglow, the so-called dark bursts (Taylor et al. 2000; Fynbo et al. 2001; Lazzati et al. 2002; Reichart & Yost 2001). In only a handful of cases we have direct evidence that the optical emission was obscured by dust, based on a

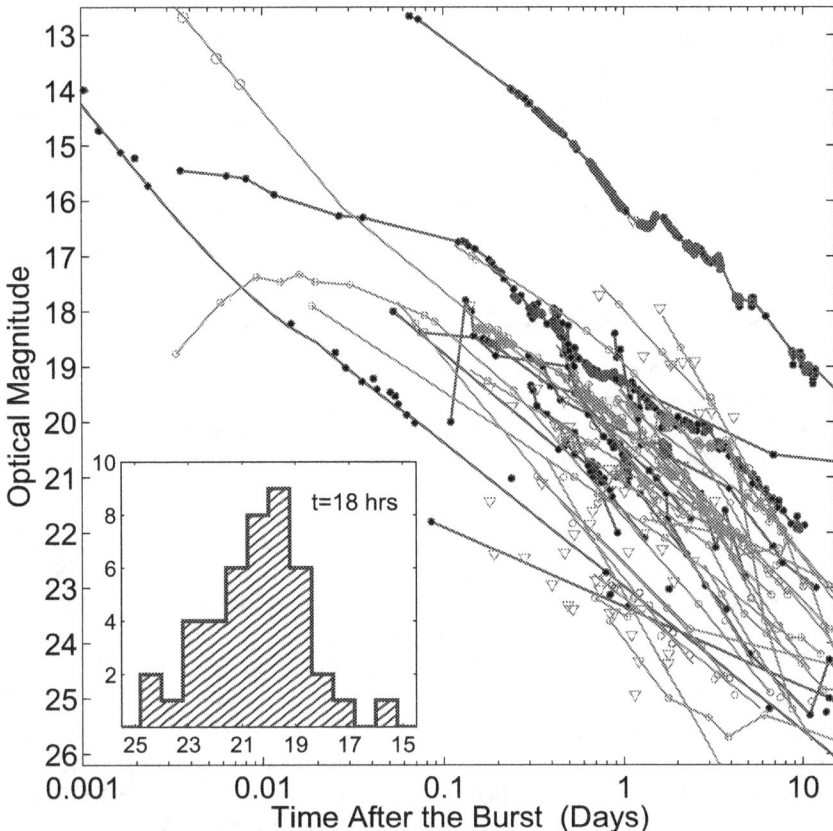

Figure 1.6: Optical light curves of 44 GRBs along with upper limits for 65 well-localized bursts. Darker shade indicates bursts localized with the *HETE* SXC for which the afterglow detection rate is about 90%. Contrary to claims that non-detections are the result of dust obscuration, the figure shows that many can simply be the result of faint afterglows. Thus, it is likely that afterglows obscured by dust comprise a small fraction of the total sample. The inset shows the wide distribution of optical magnitudes at 18 hours after the burst, extending to $R \approx 24.5$ mag.

comparison to the X-ray and/or radio emission (Djorgovski et al. 2001a; Piro et al. 2002). An alternative explanation is a high redshift, leading to absorption of the optical light in the Lyα forest. However, when host galaxies of dark bursts have been detected, the redshifts are invariably $z \sim 1$ (Djorgovski et al. 2001a; Piro et al. 2002).

Still, several authors have argued that most dark bursts are obscured by dust within their local environments (e.g., Lazzati et al. 2002; Reichart & Price 2002), and this led to the conclusion that over 50% of the cosmic star formation happens in obscured regions (Kulkarni et al. 2000; Djorgovski et al. 2001b; Ramirez-Ruiz et al. 2002; Reichart & Price 2002).

However, observations of GRB 020124 presented in Chapter 9 show that this is probably not the case. The optical emission from this burst was faint and faded relatively quickly, but the upper limit on dust extinction is $A_V < 1$ mag. Thus, with a delayed response this burst would have been classified as dark, despite the apparent lack of obscuration. A comparison to all non-detections available at this time reveals that the majority can be due to similarly faint, but non-extinguished, bursts; see Figure 1.6. If this is in fact the case, then the fraction of obscured star formation could not be easily inferred from GRBs lacking an optical afterglow.

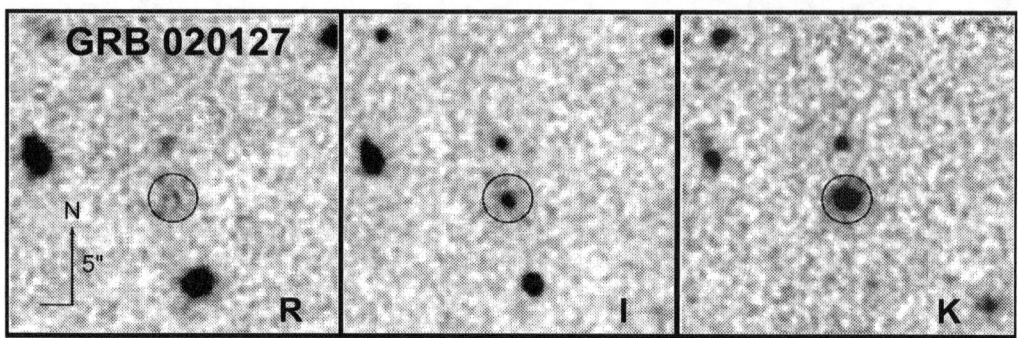

Figure 1.7: Keck optical and near-IR images of a 20 arcsec field around the position of the optically dark GRB 020127. The detection of X-ray and radio afterglows resulted in the detection of the first extremely red GRB host galaxy, with $R - K \approx 6$ mag. This host stands in direct contrast with the color distribution of nearly 40 GRB hosts for which $\langle R - K \rangle \approx 2.5$ mag.

The advent of the Soft X-ray Camera (SXC) on board the *HETE* satellite made it possible to place strict limits on the absence of optical emission, based on the accurate and rapid localizations. Surprisingly, of the thirteen bursts localized with the SXC, twelve had optical afterglows detected. The high detection rate confirms that the vast majority of past non-detections were simply due to inadequate searches. This is summarized in Figure 1.6. Thus, the fraction of truly dark bursts is $\sim 10\%$. Since GRBs are related to the formation of massive stars, and therefore explode within the stellar birth-site, this result raises three interesting possibilities: (i) Gamma-ray bursts do not occur in environments representative of the bulk of cosmic star formation, (ii) current values of the obscured fraction of star formation, $\sim 50 - 90\%$, have been severely over-estimated, or (iii) GRBs can efficiently destroy circumburst dust along the line of sight (Waxman & Draine 2000).

While the solution to this puzzle is not yet resolved (but see also §1.6.2), we can place some limits on the possibility of significant dust destruction. The radius out to which dust is destroyed depends on the the luminosity of the prompt optical/UV flash associated with the burst, $R_{\rm dest} \approx 10(L_{\rm UV}/10^{49}\,{\rm ergs}^{-1})^{1/2}$ pc (Waxman & Draine 2000). We now know (Fox et al. 2003; Li et al. 2003) that the the typical luminosities are at least an order of magnitude fainter than the first such flash detected, from GRB 990123 with an isotropic luminosity of about 10^{49} erg s^{-1} (Akerlof et al. 1999). Thus, the typical radius to which dust is destroyed is most likely less than 1 pc. In addition, since GRBs are highly collimated, dust will only be destroyed efficiently within the initial jet opening angle. As a result, when the jet begins to spread sideways (§1.5) the amount of extinction should increase. However, there is no clear evidence for such a chromatic effect following the time of the jet break in the optical/near-IR bands. Thus, while dust destruction is an inevitable process, it is not clear that it can explain the low fraction of dust-obscured bursts. Still, this process does complicate any mapping of the obscured GRB fraction to the fraction of obscured star formation.

1.6.2   What Is the Nature of GRB Host Galaxies?

The properties of GRB host galaxies impact our understanding of the progenitor systems and at the same time provide unique insight into star formation and galaxy evolution. Preliminary work, focused primarily on optical observations, has shown that GRB host galaxies span a wide range of redshifts ($z = 0.1 - 4.5$) with a peak at $z \approx 1$, and a wide range of magnitudes ($R \approx 18 - 30$ mag) with a peak at $R \approx 25$ mag. The rest-frame B-band luminosities range from about -16 to -21 mag, i.e. $\approx 0.01 - 2$ L$_*$, with the host galaxy of GRB 980326 likely having $M_B \approx -14$ mag (0.002 L$_*$). Star formation rates obtained from optical indicators (e.g., Hα) range from less than 1 M$_\odot$ yr^{-1} to about 50 M$_\odot$ yr^{-1}. Finally, low metallicities have been claimed in a few cases (e.g., Fynbo et al. 2003), but it is not clear

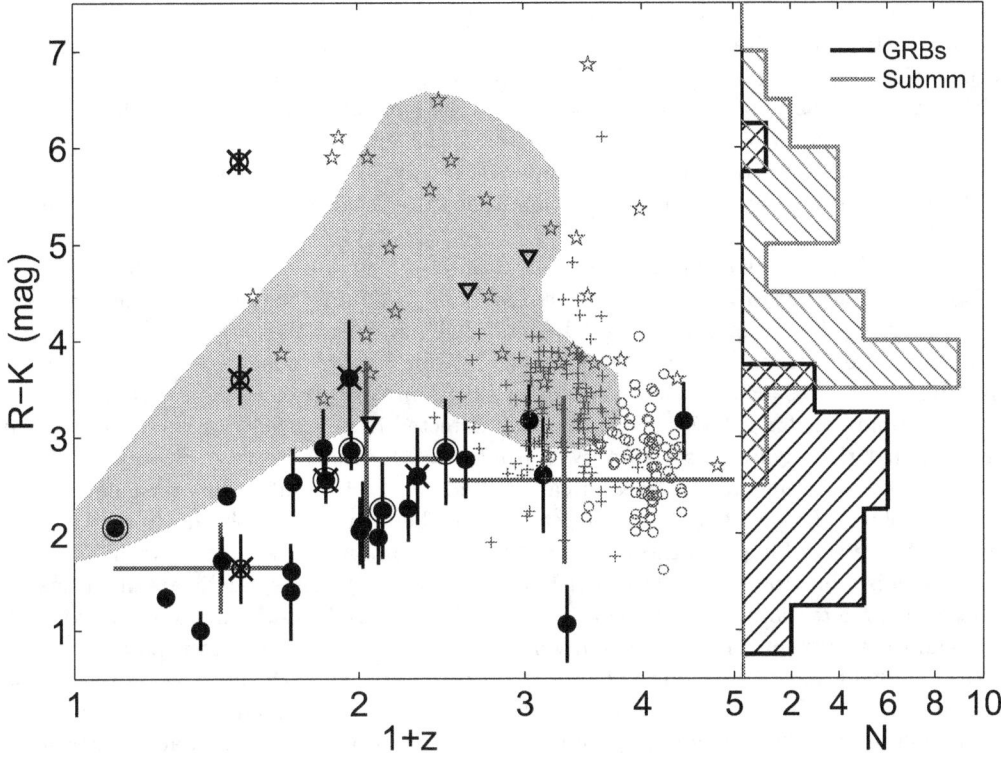

Figure 1.8: Optical/near-IR $R - K$ color plotted versus redshift for GRB hosts (filled circles), Lyman break galaxies (pluses and open circles), submillimeter galaxies (pentagrams) and K-selected galaxies in the Great Observatories Origins Deep Survey (shaded region). GRB host galaxies, including those hosting dark bursts (crossed symbols) and those detected in the radio and/or submillimeter bands (circled symbols) are significantly bluer than all other galaxy samples. This suggests that GRBs select galaxies in the early stages of their formation and starburst process.

if this is true for all GRB hosts.

While the redshift, brightness, and star formation distributions are extremely diverse, the colors of GRB host galaxies are strikingly uniform and blue. The average $R - K$ color for the sample is about 2.5 mag, with a spread of only 1 mag (Figure 1.8). It is important to note that the hosts of dust obscured GRBs are also blue, indicating that the colors of GRB hosts are not due simply to a selection against dusty (and hence red) galaxies. However, the sole exception to date, the host galaxy of GRB 020127 with $R - K \approx 6$ mag qualifying as an Extremely Red Object (Figure 1.7), did host one of the few genuine dark GRBs detected to date.

As with the GRB and afterglow phenomena themselves, a careful study of host galaxies requires multi-wavelength observations. The optical/near-IR bands provide excellent sensitivity, but they cover a small fraction of the spectral energy distribution, and are also affected by dust obscuration. However, radio emission, arising from a combination of thermal emission from HII regions and synchrotron emission from supernova remnants, is unaffected by dust and therefore provides an estimate of the total star formation rate. Similarly, far-IR emission, arising from dust-reprocessed stellar UV light, probes the obscured star formation rate. For galaxies beyond $z \sim 1$ the dust spectrum can be probed with submillimeter observations.

The initial detections of long-wavelength emission from GRB hosts occurred serendipitously. I detected radio emission from the host of GRB 980703 while monitoring the afterglow evolution (Chapter 10), while in the case of GRB 010222 we detected a persistent submillimeter source which dominated the early emission from the afterglow (Frail et al. 2002). In the former, the radio emission emanated from a region more compact than the optical host, pointing to a nuclear starburst. Furthermore, the burst position was less than 300 pc from the center of the starburst, establishing a direct connection between the GRB and the region of most intense star formation. Finally, the radio emission requires a star formation rate of about 300 $M_\odot$ yr^{-1}, compared to only 10 $M_\odot$ yr^{-1} inferred from optical spectroscopy (Djorgovski et al. 1998). A similar fraction of obscured star formation was inferred in the case of GRB 010222.

The recognition that we may be missing the bulk of the star formation, along with the possibility that GRBs preferentially select ultra-luminous starburst galaxies gave impetus to a comprehensive radio and submillimeter survey. Observations with the VLA and SCUBA reveal that about 15% of all GRB host galaxies are detectable at these wavelengths (Chapter 11), for the first time confirming observationally predictions from various cosmic star formation histories (Ramirez-Ruiz et al. 2002). However, none are as bright as the submillimeter galaxies that have been detected in blank field surveys. This may not be surprising given that these galaxies are rare, $N(> 5\,\text{mJy}) \approx 0.15$ arcmin^{-2} (e.g., Scott et al. 2002).

Despite the broad agreement with theoretical predictions, typical submillimeter galaxies have red optical/near-IR colors, $\langle R - K \rangle \approx 5$ mag, consistent with the idea of dust obscuration. As mentioned above, GRB hosts are very blue, and those detected in the submillimeter and radio with $\langle R - K \rangle \approx 2.4$ mag, are indistinguishable from the overall distribution (Figure 1.8). Thus, GRBs are intrinsically bluer since they explode preferentially in a different environment compared to submillimeter galaxies. I argue that GRBs tend to select younger starbursts in which a larger fraction of the most massive stars, which dominate the blue light, are still shining. This scenario meshes nicely with the growing realization that GRBs arise from the death of massive stars. Independent of the exact scenario, it is clear that GRB hosts detected in the submillimeter and radio represent a population of galaxies that is generally missed in current submillimeter surveys.

Since the initial starburst phase lasts a small fraction of the total lifetime of the galaxy, GRBs allow us to uniquely probe a phase of the star formation process that is generally missed in current star formation studies. The inclusion of GRB hosts may therefore significantly alter our understanding of where and under what conditions the bulk of the cosmic star formation takes place.

Part I

The Energetics of Cosmic Explosions

CHAPTER 2

A Jet Model for the Afterglow Emission from GRB 000301C[†]

E. Berger[a], R. Sari[b], D. A. Frail[c], S. R. Kulkarni[a], F. Bertoldi[d], A. B. Peck[d],
K. M. Menten[d], D. S. Shepherd[c], G. H. Moriarty-Schieven[e], G. Pooley[f], J. S. Bloom[a],
A. Diercks[a], T. J. Galama[a], & K. Hurley[g]

[a]Department of Astronomy, 105-24 California Institute of Technology, Pasadena, CA 91125, USA

[b]Theoretical Astrophysics 130-33, California Institute of Technology, Pasadena, CA 91125

[c]National Radio Astronomy Observatory, Socorro, NM 87801

[d]Max-Planck-Institut fr Radioastronomie, Auf dem Huegel 69, D-53121 Bonn, Germany

[e]Joint Astronomy Centre, 660 N. A'ohoku Place, Hilo, HI 96720

[f]Mullard Radio Astronomy Observatory, Cavendish Laboratory, Madingley Road, Cambridge CB3 0HE, England UK

[g]University of California at Berkeley, Space Sciences Laboratory, Berkeley, CA 94720

Abstract

We present broad-band radio observations of the afterglow of GRB 000301C, spanning from 1.4 to 350 GHz for the period of $3 - 130$ days after the burst. These radio data, in addition to measurements in the optical bands, suggest that the afterglow arises from a collimated outflow, i.e., a jet. To test this hypothesis in a self-consistent manner, we employ a global fit and find that a model of a jet expanding into a constant-density interstellar medium (ISM+jet) provides the best fit to the data. A model of the burst occurring in a wind-shaped circumburst medium (wind-only model) can be ruled out, and a wind+jet model provides a much poorer fit of the optical/IR data than the ISM+jet model. In addition, we present the first clear indication that the reported fluctuations in the optical/IR are achromatic, with similar amplitudes in all bands, and possibly extend into the radio regime. Using the parameters derived from the global fit, in particular a jet break time $t_{jet} \approx 7.3$ days, we infer a jet opening angle of $\theta_0 \approx 0.2$ rad; consequently, the estimate of the emitted energy in the GRB itself is reduced by a factor of 50 relative to the isotropic value, giving $E \approx 1.1 \times 10^{51}$ erg.

[†] A version of this chapter was published in *The Astrophysical Journal*, vol. 545, 56–62, (2000).

┌─ SECTION 2.1 ──┐

Introduction

└───┘

GRB 000301C is the latest afterglow to exhibit a break in its optical/IR light curves. An achromatic steepening of the light curves has been interpreted in previous events (e.g., Kulkarni et al. 1999a; Harrison et al. 1999; Stanek et al. 1999) as the signature of a jet-like outflow (Rhoads 1999; Sari et al. 1999), produced when relativistic beaming no longer "hides" the nonspherical surface, and when the ejecta undergo rapid lateral expansion. The question of whether the relativistic outflows from gamma-ray bursts (GRBs) emerge isotropically or are collimated in jets is an important one. The answer has an impact on both estimates of the GRB event rate and the total emitted energy – issues that have a direct bearing on GRB progenitor models.

An attempt by Rhoads & Fruchter (2001) to model this break using only the early-time ($t \lesssim 14$ days) optical/IR data has led to a jet interpretation of the afterglow evolution, but with certain peculiar aspects, such as a different jet break time at the R band than at the K' band. However, subsequent papers by Masetti et al. (2000a) and Sagar et al. (2000), with larger optical data sets, pointed out that there are large flux density variations ($\sim 30\%$) on timescales as short as a few hours, superposed on the overall steepening of the optical/IR light curves. While the origin of these peculiar fluctuations remains unknown, it is clear that they complicate the fitting of the optical/IR data, rendering some of the Rhoads & Fruchter (2001) results questionable.

In this paper we take a different approach. We begin by presenting radio measurements of this burst from 1.4 to 350 GHz, spanning a time range from 3 to 130 days after the burst. These radio measurements, together with the published optical/IR data, present a much more comprehensive data set, which is less susceptible to the effects of the short-timescale optical fluctuations. We then use the entire data set to fit a global, self-consistent jet model, and derive certain parameters of the GRB from this model. Finally, we explore the possibility of a wind and wind+jet global fit to the data, and compare our results with the conclusions drawn in the previous papers.

┌─ SECTION 2.2 ──┐

Observations

└───┘

Radio observations were made from 1.43 to 350 GHz at a number of facilities, including the James Clark Maxwell Telescope (JCMT[1]), the Institute for Millimeter Radioastronomy (IRAM[2]), the Owens Valley Radio Observatory Interferometer (OVRO), the Ryle Telescope, and the Very Large Array (VLA[3]). A log of these observations and the flux density measurements are summarized in Table 2.1. With the exception of IRAM, we have detailed our observing and calibration methodology in Kulkarni et al. (1999a), Kulkarni et al. (1999b), Frail et al. (2000a), and Frail et al. (2000b).

Observations at 250 GHz were made in the standard on-off mode using the Max-Planck Millimeter Bolometer (MAMBO; Kreysa et al. 1998) at the IRAM 30-m telescope on Pico Veleta, Spain. Gain calibration was performed using observations of Mars, Uranus, and Ceres. We estimate the calibration to be accurate to 15%. The source was initially observed on March 4 (Bertoldi 2000) and again on March 5 and 9 under very stable atmospheric conditions, and on March 6 with high atmospheric opacity. From March 24 to 26, the source was briefly reobserved three times for a total on+off integration time of 2000 s, but no signal was detected.

---

[1] The JCMT is operated by The Joint Astronomy Centre on behalf of the Particle Physics and Astronomy Research Council of the UK, the Netherlands Organization for Scientific Research, and the National Research Council of Canada.

[2] The Institute for Millimeter Radioastronomy (IRAM) is supported by INSU/CNRS (France), MPG (Germany), and IGN (Spain).

[3] The NRAO is a facility of the National Science Foundation operated under cooperative agreement by Associated Universities, Inc. NRAO operates the VLA.

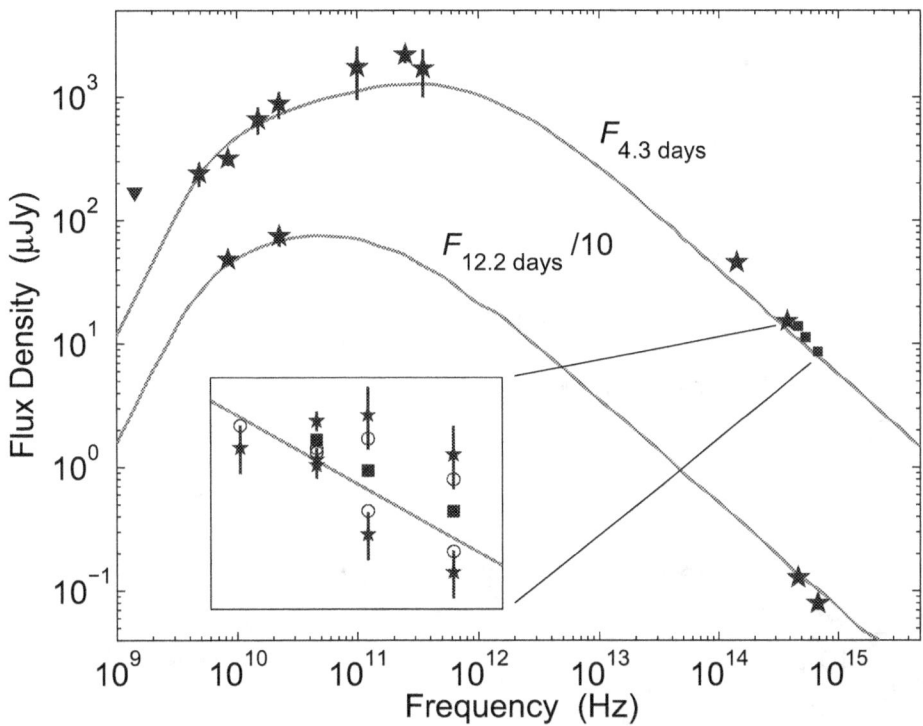

Figure 2.1: Radio to optical spectral flux distribution of GRB 000301C on 2000 March 5.66 UT ($\Delta t \approx$ 4.26 days), and 2000 March 13.58 UT ($\Delta t \approx 12.17$ days). The solid lines show the ISM+jet global fit based on a smoothed synchrotron emission spectrum (Granot et al. 1999a,b). The optical/IR data (Masetti et al. 2000a; Sagar et al. 2000; Rhoads & Fruchter 2001) are converted to Jansky flux units (Bessell & Brett 1988; Fukugita et al. 1995), and corrected for Galactic foreground extinction (Schlegel et al. 1998), with $E(B - V) = 0.053$. All data were taken within 0.5 days of the fiducial dates, and the circles show the corrections to the fiducial times, $\Delta t = 4.26$ and 12.17 days. The squares in the optical band show weighted averages of multiple measurements within 1 day of $\Delta t = 4.26$ days (see inset). The data points at 100, 250, and 350 GHz are weighted averages of the individual measurements from around day 4 (see Table 2.1). The data and fit at $\Delta t = 12.17$ days were divided by a factor of ten to avoid overlap with the $\Delta t = 4.26$ curve.

SECTION 2.3

Data

In Figure 2.1 we present broad-band spectra from March 5.66 UT ($\Delta t \approx 4.25$ days) and March 13.58 UT ($\Delta t \approx 12.17$ days). Radio lightcurves at 4.86, 8.46, 22.5, and 250 GHz from Table 2.1 are presented in Figure 2.2, while optical/IR lightcurves are shown in Figure 2.3.

The quoted uncertainties in the flux densities given in Table 2.1 report only measurement error and do not contain an estimate of the effects of interstellar scattering (ISS), which is known to be significant for radio afterglows (e.g., Frail et al. 2000c). We can get some guidance as to the expected magnitude of the ISS-induced modulation of our flux density measurements (in time and frequency) using the models developed by Taylor & Cordes (1993), Walker (1998), and Goodman (1997).

From the Galactic coordinates of GRB 000301C ($l = 48°.7$, $b = 44°.3$), we find, using the Taylor & Cordes model, that the scattering measure, in units of $10^{-3.5}$, is $SM_{-3.5} \approx 0.7$. The distance to the scattering screen, $d_{\rm scr}$, is one-half the distance through the ionized gas layer, $d_{\rm scr} = (hz/2)(\sin b)^{-1} \approx$

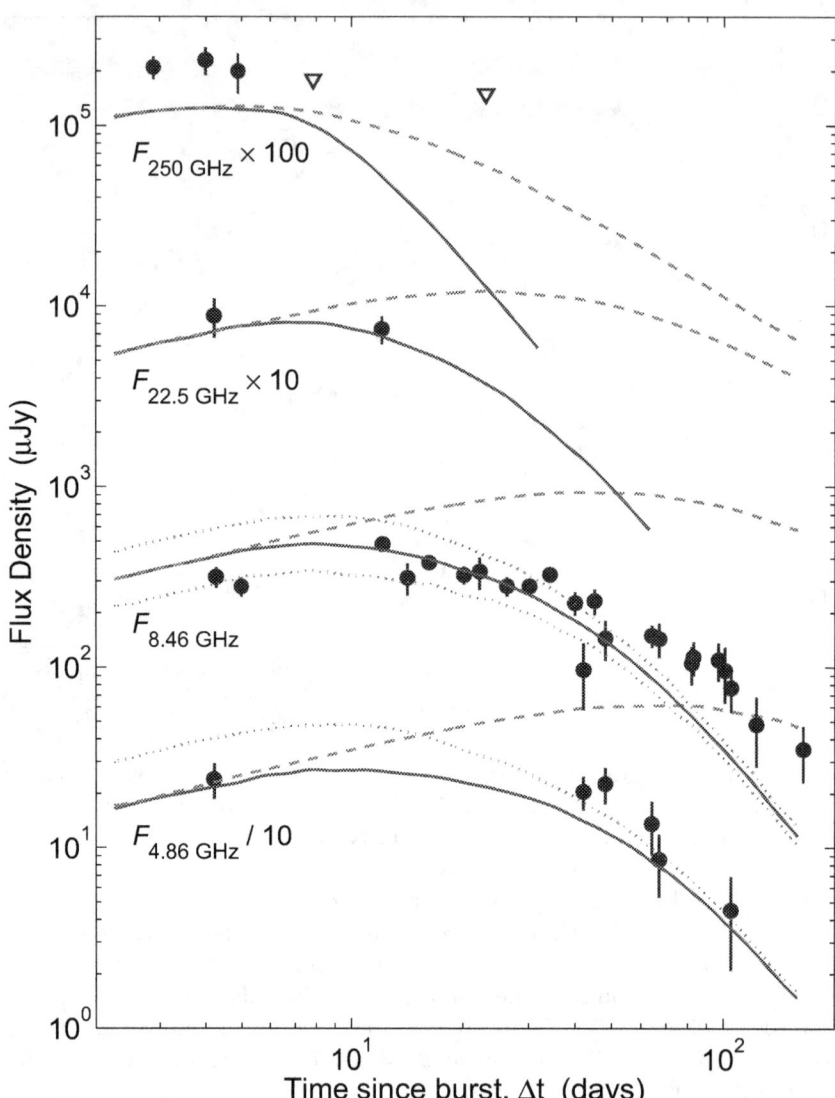

Figure 2.2: Radio lightcurves at 4.86, 8.46, 22.5 and 250 GHz. The solid lines show the ISM+jet model (§2.4). The dashed curve shows the prediction for a spherical evolution of the afterglow (ISM only). The dotted lines indicate the maximum and minimum range of flux expected from ISS (§2.3). Note that the data and fit for 4.86 GHz were divided by a factor of 10, the data and fit for 22.5 GHz were multiplied by a factor of 10, and the data and fit for 250 GHz were multiplied by a factor of 100 to avoid overlap between the four curves.

0.72 kpc, using $hz \approx 1$ kpc. From Walker's analysis, the transition frequency between weak and strong scintillation is then given by $\nu_0 = 5.9 SM_{-3.5}^{6/17} d_{\mathrm{scr}}^{5/17} \approx 4.7$ GHz. Goodman (1997) uses the same expression, but with a different normalization for the transition frequency, giving a larger value, $\nu_0 \approx 8.3$ GHz. In this section we follow Walker's analysis, and note that the numbers from Goodman will give somewhat different results.

For frequencies larger than the transition frequency, the modulation index (i.e., the rms fractional flux variation) is $m_\nu = (\nu_0/\nu)^{17/12}$, and the modulation timescale in hours is $t_\nu \approx 6.7 (d_{\mathrm{scr}}/\nu)^{1/2}$. From these equations we find that the modulation index is of the order of 0.4 at 8.46 GHz, 0.2 at 15 GHz, 0.1 at 22.5 GHz, and is negligible at higher frequencies. The modulation timescales are of the order of 2.0 hr at 8.46 GHz, 1.5 hr at 15 GHz, and 1.2 hr at 22.5 GHz. It is important to note that factor of 2 uncertainties in the scattering measure allow the modulation index to vary by $\sim 50\%$.

At these frequencies, the expansion of the fireball will begin to "quench" the ISS when the angular size of the fireball exceeds the angular size of the first Fresnel zone, $\theta_F = 8(d_{\mathrm{scr}}\nu_{\mathrm{GHz}})^{-1/2}$ μas. To describe the evolution of the source size, θ_s, with time, we have used an expanding jet model (see Frail et al. 2000b), with the factor $(E_{52}/n_1)^{1/8}$ assumed to be of order unity, which gives $\theta_s \approx 3.1(\Delta t_d/15)^{1/2}$ μas; E_{52} is the energy of the GRB in units of 10^{52} erg, n_1 is the density of the circumburst medium in units of 1 cm^{-3}, and Δt_d is the elapsed time since the burst in days. Once the source size exceeds the Fresnel size (after approximately 2 weeks at 8.46 GHz), the modulation index is reduced by a factor $(\Delta t_d/15)^{-7/12}$.

The measurements at 4.86 GHz occur near the transition frequency, and we therefore expect $m_{4.86}$ to be large, $\sim 0.65 - 1$. At 1.43 GHz, the observations were made in the strong regime of ISS, where we expect both refractive and diffractive scintillation. Point-source refractive scintillation at 1.43 GHz has a modulation index $m_{1.43,r} = (\nu/\nu_0)^{17/30} \approx 0.5$, with a timescale of $t_{1.43,r} \approx 2(\nu_0/\nu)^{11/5} \approx 1$ day. The refractive ISS is "quenched" when the angular size of the source is larger than $\theta_r = \theta_{F0}(\nu_0/\nu)^{11/5}$, where θ_{F0} is the angular size of the first Fresnel zone at $\nu_0 = 4.7$ GHz. As with weak scattering, the modulation index must be reduced by a factor $(\Delta t_d/15)^{-7/12}$ after this point. The diffractive scintillation has a modulation index $m_{1.43,d} = 1$ and a timescale $t_{1.43,d} \approx 2(\nu/\nu_0)^{6/5} \approx 0.5$ hr $\ll t_{1.43,r}$. The source can no longer be approximated by a point source when its angular size exceeds $\theta_d = \theta_{F0}(\nu/\nu_0)^{6/5}$, and correspondingly, the modulation index must be corrected by a factor $(\Delta t_d/15)^{-1/2}$.

The redshift of GRB 000301C was measured using the *Hubble Space Telescope* (HST) to be 1.95 ± 0.1 by Smette et al. (2000) and was later refined by Castro et al. (2000) using the Keck II 10-m telescope to a value of 2.0335 ± 0.0003. The combined fluence measured by the GRB detector on board the Ulysses satellite, and the X-ray/gamma-ray spectrometer (XGRS) on board the Near-Earth Asteroid Rendezvous (NEAR) satellite, in the 25-100 keV and > 100 keV bands, was 4.1×10^{-6} erg cm^{-2}. Using the cosmological parameters $\Omega_0 = 0.3$, $\Lambda_0 = 0.7$, and $H_0 = 65$ km s^{-1} Mpc^{-1}, we find that the isotropic γ-ray energy release from the GRB was $E_{\gamma,\mathrm{iso}} \approx 5.4 \times 10^{52}$ erg.

SECTION 2.4

A Self-Consistent Jet Interpretation

According to the standard, spherical GRB model, the optical light curves should obey a simple power-law decay, $F_\nu \propto t^{-\alpha}$, with α changing at most by $1/4$ as the electrons age and cool (Sari et al. 1998). From Figure 2.3, it is evident that the optical lightcurves steepen substantially ($\Delta\alpha > 1/4$) between days 7 and 8, which indicates that this burst cannot be described within this standard model of an expanding spherical blast wave. This break can be attributed to a jet-like or collimated ejecta (Rhoads 1999; Sari et al. 1999).

The jet model of GRBs predicts the time evolution of flux from the afterglow, and of the parameters $\nu_a \propto t^{-1/5}$, $\nu_m \propto t^{-2}$, and $F_{\nu,\mathrm{max}} \propto t^{-1}$, where ν_a is the self-absorption frequency, ν_m is the characteristic frequency emitted by electrons with Lorentz factor γ_m, and $F_{\nu,\mathrm{max}}$ is the observed peak flux density. This model holds for $t > t_{\mathrm{jet}}$, where t_{jet} is defined by the condition $\Gamma(t_{\mathrm{jet}}) \sim \theta_0^{-1}$. Prior to t_{jet}, the time evolution of the afterglow is described by a spherically expanding blastwave, with the scalings $\nu_a \propto$ const., $\nu_m \propto t^{-3/2}$, and $F_{\nu,\mathrm{max}} \propto$ const. In this paper we designate this model as ISM+jet. Throughout the analysis we assume that the cooling frequency, ν_c, lies above the optical band for the entire time period under discussion in this paper.

At any point in time, the spectrum is roughly given by the broken power law $F_\nu \propto \nu^2$ for $\nu < \nu_a$, $F_\nu \propto \nu^{1/3}$ for $\nu_a < \nu < \nu_m$, and $F_\nu \propto \nu^{-(p-1)/2}$ for $\nu > \nu_m$, where p is the electron power-law index. To globally fit the entire radio and optical/IR data set, we employed the smoothed form of the broken power-law synchrotron spectrum, calculated by Granot et al. (1999a) and Granot et al. (1999b). With this approach, we treat $t_{\rm jet}$, p, and the values of ν_a, ν_m, and $F_{\nu,\rm max}$ at $t = t_{\rm jet}$ as free parameters. This method forces $t_{\rm jet}$ to have the same value at all frequencies. In addition, the shape of the transition from spherical to jet evolution is described by the analytical form $F_\nu = (F_{\nu,s}^n + F_{\nu,j}^n)^{1/n}$, with n left as a free parameter. We find the following values for the burst parameters: $t_{\rm jet} = 7.3 \pm 0.5$ days, $p = 2.70 \pm 0.04$, $n = -6$, $\nu_a(t = t_{\rm jet}) = 6.8 \pm 1.8$ GHz, $\nu_m(t = t_{\rm jet}) = (3.3 \pm 0.4) \times 10^{11}$ Hz, and $F_{\nu,\rm max}(t = t_{\rm jet}) = 2.6 \pm 0.2$ mJy, where the errors are the 1σ values derived from the correlation matrix. We note that there is substantial covariance between some of the parameters, and therefore these error estimates should be treated with caution. From our fit, the asymptotic temporal decay slopes of the optical light curves are $\alpha_1 = -3(p-1)/4 = -1.28$ for $t < t_{\rm jet}$, and $\alpha_2 = -p = -2.70$ for $t > t_{\rm jet}$. The fits are shown in Figures 2.1–2.3.

The total value of χ^2 for the global fit is poor. We obtain $\chi^2 = 670$ for 140 degrees of freedom. The bulk of this value, 550, comes from the 102 optical data points, and is the result of the observed fluctuations, which are not accounted for by our model. The radio data contribute a value of 120 to χ^2 for 43 data points. This is the result of scintillation, and the observed late-time flattening of the 8.46 GHz lightcurve. If we increase the errors to accommodate the expected level of scintillation (see §2.3), we obtain a good fit with $\chi^2_{\rm radio} = 45/38$ degrees of freedom.

From Figure 2.1, it is clear that the global fit accurately describes the broad-band spectra on days 4.26 ($t < t_{\rm jet}$) and 12.17 ($t > t_{\rm jet}$), with a single value of $p = 2.70$, which rules out the possibility that the steepening of the lightcurves at $t_{\rm jet}$ is the result of a time-varying p.

Trying to model the data using the approach outlined above, but for a wind-shaped circumburst medium, results in a poor description of the data, because the wind model does not exhibit a break, although one is clearly seen in the optical data. As a result, the model fit is too low at early times, and too high at late times relative to the data (see inset in Figure 2.3). The value of χ^2 for the wind model relative to the ISM+jet model described above is $\chi^2{\rm wind}/\chi^2_{\rm ISM+jet} \approx 4$. Therefore, the wind model can be ruled out as a description of the afterglow of GRB 000301C.

A jet evolution combined with a wind-shaped circumburst medium provides a more reasonable fit than a wind-only model. The wind evolution of the fireball will only be manifested for $t < t_{\rm jet}$, since once $\Gamma(t_{\rm jet}) \approx \theta_0^{-1}$ the jet will expand sideways and appear to observers as if it were expanding into a constant-density medium (Chevalier & Li 1999, 2000; Livio & Waxman 2000; Kumar & Piran 2000). The resulting parameters from such a fit differ considerably from the parameters for the ISM+jet model quoted above, and the relative value of χ^2 between the two models is $\chi^2{\rm wind + jet}/\chi^2_{\rm ISM+jet} \approx 2$. This model suffers from a serious drawback in its description of the optical/IR lightcurves. Because the predicted decay of these lightcurves prior to $t_{\rm jet}$ is steeper than in the ISM+jet model, the model fit, from 2 days after the burst up to the break time, is too low relative to the data (see inset in Figure 2.3).

It is important to note that in a recent paper, Kumar & Piran (2000) suggested that the steepening of the lightcurves when the jet geometry of the outflow becomes manifested is completed over 1 decade in observer time in the case of the ISM+jet scenario, and over $2-4$ decades in observer time in the wind+jet scenario. We can estimate the transition time, δt, by comparing the smooth-transition lightcurves to the asymptotic slopes at times much larger and smaller than $t_{\rm jet}$ (i.e., the same light curves but with a sharp transition). We find that the maximum deviation between the two curves, which occurs at $t_{\rm jet}$, is $\sim 10\%$. If we therefore define the transition time as the period during which the sharp curves deviate by more than 1% (or 5%) from the smooth curves, then we find that the transition time for GRB 000301C is approximately 10 (or 4) days, which gives $\delta t/t_{\rm jet} \approx 1$. This transition time is clearly inconsistent with the extremely gradual steepening in a wind-shaped circumburst medium, but is consistent with the expected transition time in the ISM+jet case. However, Kumar & Piran (2000) claim that the expected change in the power-law index due to the jet break is $\Delta\alpha \sim 0.7$, while the observed steepening in this case is $\Delta\alpha \approx 1.4$. A similar behavior in the afterglow emission from GRB 990510 was explained as the result of the passage of ν_c and ν_m through the optical bands at $t \sim t_{\rm jet}$. In the case of GRB 000301C, however, we expect ν_m to cross the optical band at $t \approx 0.05$ days $\ll t_{\rm jet}$. In the context of this analysis,

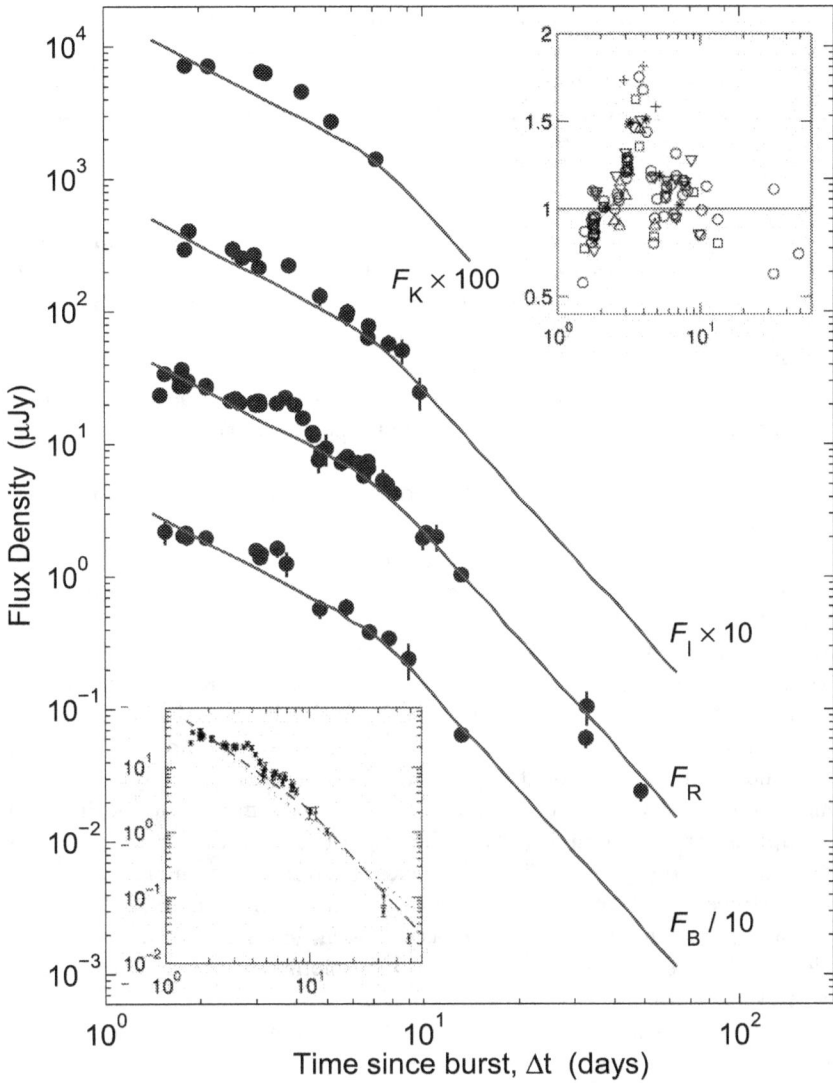

Figure 2.3: Optical/near-IR lightcurves of GRB 000301C. Following Masetti et al. (2000a), we added a 5% systematic uncertainty in quadrature to all optical measurements to account for discrepancies between the different telescopes and instruments. The solid lines show the ISM+jet global fit. In the top right inset are plotted the data points divided by the respective model fit for all bands (circles, squares, stars, triangles, inverted triangles, and plus signs: R, B, K', V, I, and 250 GHz bands, respectively). The short-timescale fluctuations are clearly achromatic and with a comparable amplitude in all bands, possibly spanning from optical to radio. The inset in the bottom left portion of the figure shows the global fits based on the wind-only (dotted line), and wind+jet (dashed line) models overlaid on the R-band data. The steeper decline predicted for a fireball expanding into a wind-shaped medium results in a much poorer fit relative to the ISM+jet model (§2.4). Note that the data and fit for the B band were divided by a factor of 10, the data and fit for the I band were multiplied by a factor of 10, and the data and fit for the K' band were multiplied by a factor of 100 to avoid overlap between the four curves.

ν_c does not cross the optical band at all, and in a fit in which ν_c is left as a free parameter, it is expected to cross the optical band at $t \sim 10^{-3}$ days (see §2.5).

The global fitting approach has several advantages over fitting each component of the data set independently. For example, the K' data are only available up to day 7.18 ($\lesssim t_{\rm jet}$) after the burst. Therefore, by fitting them independently of the other optical bands and the radio data, we cannot find $t_{\rm jet}$ if it is indeed after 7 days. Moreover, as we can see from Figure 2.3, there is an additional process that superposes achromatic fluctuations with an overall rise and decline centered between days 3 and 4, on top of the smoothly decaying optical emission (see inset in Figure 2.3), then fitting the K' data independently will confuse this behavior with the jet break. This explains the result of Rhoads & Fruchter (2001) of $t_{\rm jet,K'} \sim 3$ days. It is worth noting that fitting the available R-band data from before day 8 by itself gives a value of $t_{\rm jet,R} \sim 3.5$ days $\sim t_{\rm jet,K'}$.

Simultaneous fitting of the entire data set makes it possible to study the overall behavior of the fireball regardless of any additional sources of fluctuations, because the large range in frequency and time of the data reduces the influence of such fluctuations. Remarkably, using this global fit with only the radio data, ignoring the optical observations, we obtain $t_{\rm jet,radio} \approx 7.7$ days $\sim t_{\rm jet}$. Thus, the radio data serve to support the jet model, and provide an additional estimate for the jet break time, independent of the somewhat ambiguous optical data.

From the global fit, we find the first self-consistent indication that the short-timescale optical fluctuations are achromatic, even in the K' band (see inset in Figure 2.3). By simply dividing the B, V, R, I and K' data by the values from the global fit, we find that the fluctuations happen simultaneously and with similar amplitudes in all bands. Moreover, the overall structure of the fluctuations is a sharp rise and decline centered on day 4, and with an overall width of 3.5 days, which gives $\delta t/t \sim 1$, where δt is the width of the bump. The optical/IR data start at day 1.5 lower by $25 - 50\%$ than the model fit, then rise to a peak level of $50 - 75\%$ relative to the model at day 4, and drop to the predicted level at about day 6, at which point they follow the predicted decline of the ISM+jet model.

It is interesting to note that the 250 GHz data, which are not affected by ISS-induced fluctuations, also show a peak amplitude approximately 70% higher than the model fit around day 4 (see Figure 2.2 and inset in Figure 2.3). At the lower radio frequencies, there are not enough data points to discern a similar behavior. Moreover, at these frequencies it would have been difficult to disentangle such fluctuations from ISS-induced fluctuations in any case. The large range in frequency of this achromatic fluctuation, coupled with the similar level of absolute deviation from the model fit, suggests that it is the result of a real physical process.

It is possible to explain this fluctuation as the result of a nonuniform ambient density. The value of ν_m is independent of the ambient medium density, and since $F_{\nu,\max} \propto n_1^{1/2}$, we expect the flux at frequencies larger than ν_a to vary achromatically, and with the same amplitude, $F_\nu \propto n_1^{1/2}$. For frequencies lower than ν_a we must take into account the density dependence, $\nu_a \propto n_1^{3/5}$, so that the flux will vary according to $F_\nu \propto F_{\nu,\max}\nu_a^{-2} \propto n_1^{-7/10}$. This means that for frequencies lower than ~ 7 GHz, we actually expect the flux to fluctuate downward at the same time that it fluctuates upward at higher frequencies. In practice, we do not have enough data around this time to confirm this behavior, but we do note that the two data points at 8.46 GHz from around day 4 exhibit a lower flux density level than expected from the fit. This discrepancy, however, can also be due to ISS. In order to match the observed peak amplitude of the optical fluctuation, of order 80%, the ambient density must vary by about a factor of 3.

Using the value of $t_{\rm jet}$ from our global fit, we can calculate the jet opening angle, θ_0, from the equation:

$$\theta_0 \approx 0.05 t_{\rm j,hr}^{3/8}(1+z)^{-3/8}(n_1/E_{52})^{1/8} \qquad (2.1)$$

(Sari et al. 1999; Livio & Waxman 2000), where E_{52} is the isotropic energy release, which can be roughly estimated from the observed fluence; using the equations from Rhoads (1999) results in a smaller opening angle. From this equation, we calculate a value of $\theta_0 \approx 0.2 n_1^{1/8}$ rad. This means that the actual energy release from GRB 000301C is reduced by a factor of 50 relative to the isotropic value, $E_{\gamma,\mathrm{iso}} \approx 5.4 \times 10^{52}$ erg, which gives $E_\gamma = 1.1 \times 10^{51} n_1^{1/4}$ erg.

SECTION 2.5

Conclusions

The afterglow emission from GRB 000301C can be well described in the framework of the jet model of GRBs. Global fitting of the radio and optical data allows us to calculate the values of p, t_{jet}, the time evolution of ν_a, ν_m and $F_{\nu,\mathrm{max}}$, and the shape of the transition to jet evolution in a self-consistent manner. Within this approach, the proposed discrepancy between the behaviors of the R- and K'-band lightcurves, suggested by Rhoads & Fruchter (2001), is explained as the result of the lack of data for $t > 7.18$ days ($\lesssim t_{\mathrm{jet}}$) at K', and the existence of achromatic substructure from fluctuations in the optical/IR, and possibly the radio regime. The value for the break time from the global, self-consistent approach we have used is $t_{\mathrm{jet}} = 7.3$ days at all frequencies.

The long-lived radio emission from the burst, spanning a large range in frequency and time, plays a significant role in our ability to extract the time evolution of ν_a, ν_m and $F_{\nu,\mathrm{max}}$ from the data. In the case of this GRB in particular, the large range in frequency and time is crucial, since it serves to reduce the effects of unexplained deviations from the simple theory, such as the short-timescale fluctuations in the optical bands, on the overall evolution of the fireball.

We end with some general remarks about the fit in the case in which ν_c is not constrained to lie above the optical band. If we just add ν_c as an additional free parameter in the fit, we find that the best-fit value for the cooling frequency at t_{jet} is $\nu_c \approx 5 \times 10^{14}$ Hz, while the best-fit values for all other parameters are relatively unchanged (i.e., within 2σ of the values given in §2.4). This value of ν_c indicates that the cooling frequency crosses the optical bands approximately 2 days after the burst. However, the resulting modest steepening of $\Delta\alpha = 1/4$ is overshadowed by the much larger scale overall fluctuation in the optical bands. Using a different approach, in which we fix the value of ν_c and leave all other parameters to vary freely, we find that for all values of ν_c, the value of ν_m is lower than ν_c at $t = t_{\mathrm{jet}}$. Finally, we note that in both cases – a fixed or freely varying ν_c – the value of χ^2 is similar to the value for the analysis in §2.4.

Research at the Owens Valley Radio Observatory is supported by the National Science Foundation through NSF grant AST 96-13717. K.H. acknowledges Ulysses support under JPL contract 958056 and NEAR support under NAG5-9503.

Table 2.1. Radio and Submillimeter Observations of GRB 000301C

Epoch (UT)	Δt (days)	Telescope	ν_0 (GHz)	$S \pm \sigma$ (μJy)
2000 Mar 4.29	2.88	IRAM 30-m	250	2100 ± 300
2000 Mar 4.75	3.34	JCMT	350	3736 ± 3700
2000 Mar 4.98	3.57	Ryle	15.0	660 ± 160
2000 Mar 5.41	4.00	IRAM 30-m	250	2300 ± 400
2000 Mar 5.53	4.12	JCMT	350	2660 ± 1480
2000 Mar 5.57	4.16	OVRO	100	2850 ± 950
2000 Mar 5.67	4.26	VLA	1.43	11 ± 79
2000 Mar 5.67	4.26	VLA	4.86	240 ± 53
2000 Mar 5.67	4.26	VLA	8.46	316 ± 41
2000 Mar 5.67	4.26	VLA	22.5	884 ± 216
2000 Mar 6.29	4.88	IRAM 30-m	250	2000 ± 500
2000 Mar 6.39	4.98	VLA	8.46	289 ± 34
2000 Mar 6.50	5.09	JCMT	350	1483 ± 1043
2000 Mar 6.57	5.16	OVRO	100	-99 ± 1500
2000 Mar 9.25	7.84	IRAM 30-m	250	400 ± 600
2000 Mar 10.21	8.80	Ryle	15.0	480 ± 300
2000 Mar 13.58	12.17	VLA	8.46	483 ± 26
2000 Mar 13.58	12.17	VLA	22.5	748 ± 132
2000 Mar 15.58	14.17	VLA	8.46	312 ± 62
2000 Mar 17.61	16.20	VLA	8.46	380 ± 29
2000 Mar 21.52	20.12	VLA	8.46	324 ± 36
2000 Mar 23.55	22.14	VLA	8.46	338 ± 69
2000 Mar 24.29	22.88	IRAM 30-m	250	-300 ± 500
2000 Mar 27.55	26.14	VLA	8.46	281 ± 34
2000 Mar 31.53	30.12	VLA	8.46	281 ± 25
2000 Apr 4.59	34.18	VLA	8.46	325 ± 27
2000 Apr 10.36	39.95	VLA	8.46	227 ± 33
2000 Apr 12.47	42.06	VLA	4.86	210 ± 43
2000 Apr 12.47	42.06	VLA	8.46	91 ± 38
2000 Apr 15.43	45.02	VLA	8.46	233 ± 37
2000 Apr 18.47	48.06	VLA	4.86	226 ± 51
2000 Apr 18.47	48.06	VLA	8.46	145 ± 36
2000 May 4.49	64.13	VLA	4.86	136 ± 45
2000 May 4.49	64.13	VLA	8.46	150 ± 20
2000 May 7.50	67.09	VLA	4.86	85 ± 33
2000 May 7.50	67.09	VLA	8.46	144 ± 31
2000 May 22.45	82.04	VLA	8.46	105 ± 25
2000 May 23.45	83.04	VLA	8.46	114 ± 24
2000 Jun 6.40	96.99	VLA	8.46	110 ± 26
2000 Jun 10.31	100.90	VLA	8.46	96 ± 33
2000 Jun 14.26	104.85	VLA	4.86	45 ± 24
2000 Jun 14.29	104.88	VLA	8.46	77 ± 21
2000 Jul 2.06	122.65	VLA	8.46	48 ± 20

Note. — The columns are (left to right), (1) UT date of the start of each observation, (2) time elapsed since the γ-ray burst, (3) telescope name, (4) observing frequency, and (5) peak flux density at the best fit position of the radio transient, with the error given as the root mean square noise on the image. The JCMT observations did not detect the source at each epoch individually, but by averaging the 3.875 hr of integration over the three epochs, we obtain a 2.5σ detection of 1.70 ± 0.71 mJy.

CHAPTER 3

GRB 000418: A Hidden Jet Revealed[†]

E. Berger[a], A. Diercks[a], D. A. Frail[b], S. R. Kulkarni[a], J. S. Bloom[a], R. Sari[c],
J. Halpern[d], N. Mirabal[d], G. B. Taylor[b], K. Hurley[e], G. Pooley[f], K. M. Becker[g],
R. M. Wagner[h], D. M. Terndrup[h], T. Statler[i], D. R. Wik[i], E. Mazets[j], & T. Cline[k]

[a]Department of Astronomy, 105-24 California Institute of Technology, Pasadena, CA 91125

[b]National Radio Astronomy Observatory, P. O. Box 0, Socorro, NM 87801

[c]California Institute of Technology, Theoretical Astrophysics 103-33, Pasadena, CA 91125

[d]Astronomy Department, Columbia University 550 West 120th St., New York, NY 10027

[e]University of California, Berkeley, Space Sciences Laboratory, Berkeley, CA 94720-7450

[f]Mullard Radio Astronomy Observatory, Cavendish Laboratory, Madingley Road, Cambridge CB3 0HE

[g]Department of Physics, Oberlin College Oberlin, OH 44074

[h]Ohio State University, Department of Astronomy, Columbus, OH, 43210

[i]Ohio University, Department of Physics and Astronomy, Athens, OH, 45701

[j]Ioffe Physico-Technical Institute, St. Petersburg, 194021 Russia

[k]NASA Goddard Space Flight Center, Code 661, Greenbelt, MD 20771

Abstract

We report on optical, near-infrared and centimeter radio observations of GRB 000418 which allow us to follow the evolution of the afterglow from 2 to 200 days after the γ-ray burst. In modeling these broadband data, we find that an isotropic explosion in a constant density medium is unable to simultaneously fit both the radio and optical data. However, a jet-like outflow into either a constant density or windstratified medium with an opening angle of 10-20° provides a good description of the data. The evidence in favor of a jet interpretation is based on the behavior of the radio light curves, since the expected jet break is masked at optical wavelengths by the light of the host galaxy. We also find evidence for extinction, presumably arising from within the host galaxy, with A_V^{host}=0.4 mag, and host flux densities of $F_R = 1.1$ μJy and $F_K = 1.7$ μJy. These values supercede previous work on this burst due to the availability of a broad-band data set allowing a global fitting approach. A model in which the GRB explodes isotropically into a wind-stratified circumburst medium cannot be ruled out by these data. However, in examining a sample of other bursts (e.g., GRB 990510, GRB 000301C) we favor the jet interpretation for GRB 000418.

[†] A version of this chapter was published in *The Astrophysical Journal*, vol. 556, 556-561, (2001).

┌─ SECTION 3.1 ──┐
| |
| # Introduction |
| |
└──┘

GRB 000418 was detected on 18 April 2000, at 09:53:10 UT by the *Ulysses*, *KONUS-Wind* and *NEAR* spacecraft, which are part of the third interplanetary network (IPN). The event lasted ~30 s, and a re-analysis of the early Ulysses data (Hurley et al. 2000) gives a fluence of 4.7×10^{-6} erg cm^{-2} in the 25-100 keV band. A fit to the total photon spectrum from the KONUS data in the energy range $15 - 1000$ keV gives a fluence of 2×10^{-5} erg cm^{-2}. Intersecting IPN annuli resulted in a 35 arcmin^2 error box, in which Klose et al. (2000a) identified a variable near-infrared (NIR) source. The early R-band light curve of this source was described by Mirabal et al. (2000) as having a power-law decay $t^{-0.84}$, typical for optical afterglows. The redshift for the host galaxy of $z \simeq 1.119$ was measured by Bloom et al. (2000) from an [OII] emission line doublet. Assuming cosmological parameters of Ω_M=0.3, Λ_0=0.7 and H$_0$=65 km s^{-1} Mpc^{-1}, this redshift corresponds to a luminosity distance d$_L$ = 2.5×10^{28} cm and gives an implied isotropic γ-ray energy release of E$_\gamma = 1.7 \times 10^{52}$ erg.

Klose et al. (2000b) have recently summarized optical/NIR data observations of GRB 000418. In this paper we present additional optical/NIR data and a complete set of radio observations between 1.4 GHz and 22 GHz, from 10 to 200 days after the burst. We use this broad band data set to fit several models, deriving the physical parameters of the system.

┌─ SECTION 3.2 ──┐
| |
| # Observations |
| |
└──┘

3.2.1 Optical Observations

In Table 3.1 we present deep optical photometry obtained at Palomar, Keck[1] , and MDM observatories covering six weeks following the GRB as well as data from the extant literature.

All of the optical data was overscan corrected, flat-fielded, and combined in the usual manner using IRAF (Tody 1993). PSF-fitting photometry was performed relative to several local comparison stars measured by Henden (2000) using DoPhot (Schechter et al. 1993). Short exposures of the field in each band were used to transfer the photometry (Henden 2000) to several fainter stars in the field. Several of the Keck+ESI measurements, and the Palomar 200" measurement were made in Gunn-r and Gunn-i respectively and were calibrated by transforming the local comparison stars to the Gunn system using standard transformations (Wade et al. 1979; Jorgensen 1994). We add an additional 5% uncertainty in quadrature with the statistical uncertainties to reflected the inherent imprecision in these transformations.

The Ks-band image of the field was obtained on the Keck I Telescope on Manua Kea, Hawaii with the Near Infrared Camera (NIRC; Matthews & Soifer 1994). We obtained a total of 63 one-minute exposures which we reduced and combined with the `IRAF/DIMSUM` package modified by D. Kaplan. There was significant cloud and cirrus cover and so the night was not photometric.

The HST STIS/Clear image was obtained on 4 June 2000 UT as part of the TOO program # 8189 (P.I. A. Fruchter) and made public on 2 September 2000 UT. Five images of 500 s each were obtained which we combined using the `IRAF/DITHER` task. The final plate scale is 25 milliarcsec pixel^{-1}.

We corrected all optical measurements in Table 3.1 for a Galactic foreground reddening of $E(B-V) = 0.032$ (Schlegel et al. 1998) at the position of the burst $(l, b) = (261.16, 80.78)$ before converting to flux units Bessell & Brett (1988); Fukugita et al. (1995) assuming R$_V$=3.1.

---

[1] The W. M. Keck Observatory is operated by the California Association for Research in Astronomy, a scientific partnership among California Institute of Technology, the University of California and the National Aeronautics and Space Administration.

3.2.2   Radio Observations

Radio observations were undertaken at a frequency of 15 GHz with the Ryle Telescope. All other frequencies were observed with either the NRAO[2] Very Large Array (VLA) or the Very Long Baseline Array (VLBA). A log of these observations can be found in Table 3.2. The data acquisition and calibration for the Ryle and the VLA were straightforward (see Frail et al. 2000a for details).

The single VLBA observation was carried out at 8.35 GHz with a total bandwidth of 64 MHz in a single polarization using 2 bit sampling for additional sensitivity. The nearby ($<1.3°$) calibrator J1224+2122 was observed every 3 minutes for delay, rate and phase calibration. Amplitude calibration was obtained by measurements of the system temperature in the standard way. The coordinates for GRB 000418 derived from the VLBA detection are (epoch J2000) $\alpha = 12^h25^m19.2840^s$ ($\pm0.015^s$) $\delta = +20°06'11.141''$ ($\pm0.001''$).

SECTION 3.3

The Optical Light Curve and Host Galaxy

In Figure 3.1 we display the R and K-band light curves constructed from measurements in Table 3.1. The pronounced flattening of the R-band light curve at late times is reasonably attributed to the optical afterglow fading below the brightness of the underlying host galaxy. A noise-weighted least squares fit was made to the data of the form $f_R = f_o t_o^\alpha + f_{host}$ for which we derive $f_o = 23.4 \pm 2.1$ μJy, $\alpha_o = -1.41 \pm 0.08$ and $f_{host} = 1.08 \pm 0.06$ μJy with a reduced $\chi_r^2 = 0.94$. Our inferred R-band magnitude for the host galaxy $R_{host} = 23.66 \pm 0.06$ is nearly identical to that obtained from a similar analysis by Klose et al. (2000b). In order to estimate the effect of the host in other optical bands we scaled R_{host} for GRB 000418 to a spectrum of the host galaxy of GRB 980703 (Bloom et al. (1998a)) ($z = 0.966$) whose magnitude was measured in seven broad-band colors (B, V, R, I, J, H, and K). Our results indicate that 50-100% of the flux in some bands is due to the host galaxy after the first 10 days. Therefore, for the afterglow modeling in §3.5 we chose not to include the late-time measurements of GRB 000418 in the B, V, and Gunn-i bands.

SECTION 3.4

The Radio Light Curves

In Figure 3.1 we display the radio light curves at 4.86, 8.46, 15 and 22 GHz. To first order all four frequencies show a maximum near 1 mJy on a time scale of 10 to 20 days. There is no discernible rising trend at any frequency. This is most clear at 8.46 GHz, where beginning 10 days after the burst, the light curve undergoes a steady decline, fading from 1 mJy to 0.1 mJy over a six month period. The temporal slope of the 8.46 GHz light curve after the first two months $\alpha_{\rm rad} = -1.37 \pm 0.10$ ($\chi_r^2 = 1.4$) is similar to the optical R-band curve $\alpha_{\rm opt} = -1.41 \pm 0.08$.

Superimposed on this secular decrease, there exists point-to-point variability of order 50%, especially in the early measurements. We attribute these variations to interstellar scintillation (ISS). The method by which we estimate the magnitude of the intensity fluctuations induced by ISS as a function of frequency and time is described in full by Berger et al. (2000). Briefly, we estimate the magnitude of scattering with the model of Taylor & Cordes (1993), and use this to calculate the transition frequency ν_0 between weak and strong scattering using Walker 1998. The normalizations used in Goodman (1997) give slightly larger values of ν_0.

In the direction toward GRB 000418 we derive $\nu_0 \simeq 3.6$ GHz and therefore most of our measurements were taken in the weak ISS regime. In this case the modulation scales as $\nu^{-17/12}$, with a maximum of 65% expected at 4.86 GHz and 30% at 8.46 GHz. At 15 GHz and 22 GHz we estimate that the

---

[2] The NRAO is a facility of the National Science Foundation operated under cooperative agreement by Associated Universities, Inc. NRAO operates the VLA and the VLBA

ISS-induced fluctuations are only a fraction of the instrumental noise. The expansion of the fireball will eventually quench ISS when the angular size of the fireball exceeds the angular size of the first Fresnel zone at the distance of the scattering screen. The fireball size, and hence the quenching timescale, is model-dependent, and we use the derived total energy and density from the global fits (see §3.5 below) to estimate this time for each model. For example, in a simple spherical fireball this occurs after 15 days at 4.86 GHz and 10 days at 8.46 GHz, and thereafter the modulation indices decline as $t^{-35/48}$. We note that the observed fluctuations at 4.86 and 8.46 GHz conform to the predicted level of ISS, but that the measurements at 8.46 GHz from around 50 days after the burst deviate by a factor of three from the predicted ISS level.

In addition, we use the scintillation pattern to estimate the true χ_r^2 for each model, by adding in quadrature to the instrumental noise an additional ISS-induced uncertainty, $\sigma_{\rm ISS} = m_p F_{\nu,\rm model}$, where m_p and $F_{\nu,\rm model}$ are the modulation index and model flux density at frequency ν, respectively (Berger et al. 2000).

SECTION 3.5

Global Model Fits

The optical and radio data presented here have allowed us to track the evolution of the GRB 000418 afterglow from 2 to 200 days after the burst. With careful modeling of the light curves, it should be possible to infer the physical parameters of the blast wave and thereby gain some insight into the nature of GRB progenitors. In particular, the hydrodynamic evolution of the shock is governed by the energy of the explosion, the geometry of the expanding ejecta shock and the type of environment into which the GRB explodes (Sari et al. 1998; Wijers & Galama 1999; Chevalier & Li 1999; Panaitescu & Kumar 2000). We consider four basic models: a spherical explosion or collimated ejecta (i.e., jets) in both a constant density medium and a wind-blown medium.

The starting point for any afterglow interpretation is the cosmological fireball model (e.g., Meszaros & Rees 1997; Waxman 1997). A point explosion of energy E_0 expands relativistically into the surrounding medium (with density $\rho \propto r^{-s}$, where $s = 0$ for constant density ISM and $s = 2$ for a wind) and the shock produced as a result of this interaction is a site for particle acceleration. The distribution of electrons is assumed to be a power-law of index p, and the fraction of the shock energy available for the electrons and the magnetic field is ϵ_e and ϵ_B, respectively. The values of these three quantities (p, ϵ_e and ϵ_B) are determined by the physics of the shock and the process of particle acceleration and in the absence of detailed understanding are taken to be constant with time.

The instantaneous broad-band synchrotron spectrum can be uniquely specified by the three characteristic frequencies ν_a, ν_m, and ν_c, (i.e., synchrotron self-absorption, synchrotron peak, and cooling), the peak flux density f_m, and p. For this work we adopt the smooth spectral shape as given by Granot et al. (1999a), rather than the piecewise, broken power-law spectrum used by other authors (e.g., Wijers & Galama 1999). The evolution of the spectrum (and thus the time dependence of ν_a, ν_m, ν_c and f_m) is governed by the geometry of the explosion (spherical or a collimated into a jet-like outflow), and the properties of the external environment (constant density or a radial density profile). Our approach is to adopt a model (sphere, wind, jet, etc.) and solve for the above spectral parameters using the entire optical and radio data set. The advantages and details of global model fitting are discussed by Berger et al. (2000).

The simplest model is a spherically symmetric explosion in a constant density medium (*ISM*: Sari et al. 1998). The total χ_r^2 for this model (see Table 3.3) gives a highly unsatisfactory fit to the data. On close inspection (Figure 3.1) we find that the model systematically underpredicts the optical flux. Adding extinction from the host galaxy only makes this worse. The fundamental difficulty with the *ISM* model is that it predicts $f_m = constant$, independent of frequency. In this case, since it is the radio data that is responsible for defining the peak of spectrum, it results in a value of f_m that is too low at higher frequencies.

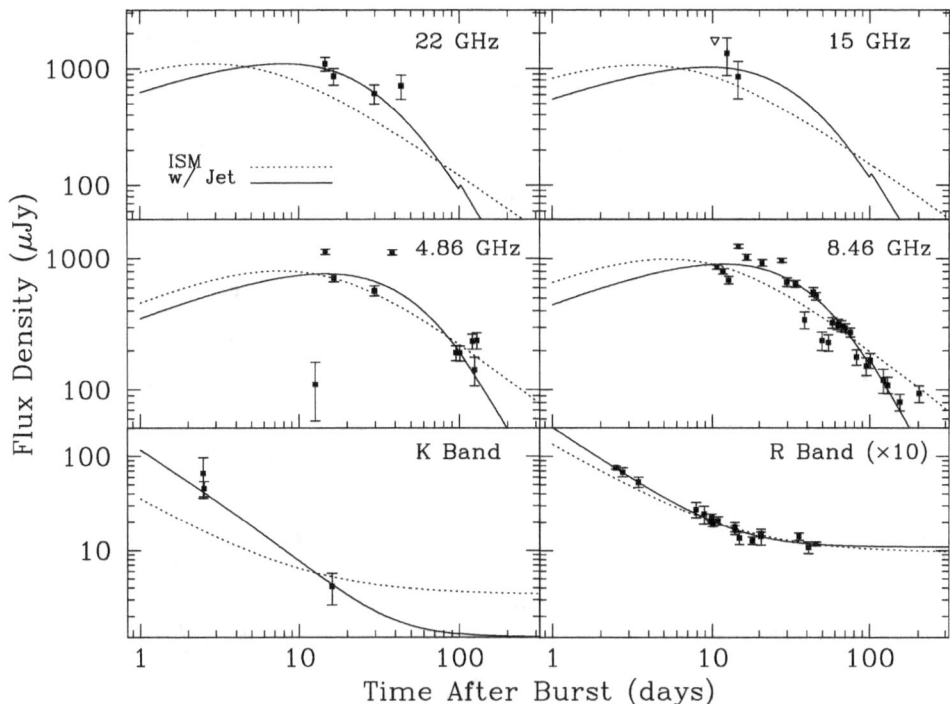

Figure 3.1: Radio and optical light curves for GRB 000418. The observing frequency (or band) is shown in the upper right corner of each panel. Optical magnitudes were first corrected for Galactic foreground reddening before converting to flux units. For display purposes the R band flux densities have been increased by a factor of 10. The 8.46 GHz measurements on August 25 and September 18 are 3-epoch averages taken over a period of 7 days and 15 days, respectively. The dotted and solid lines are light curves assuming an isotropic explosion in a constant density medium (ISM) and one in which the ejecta are collimated with opening angle θ_j (ISM+Jet), respectively. They were derived from a global fit to the entire broad-band dataset. See text for more details.

To obtain better fits to the joint optical and radio data sets we look to models for which f_m is time-dependent. One such model is a collimated outflow into a medium with uniform density (*ISM+Jet*: Rhoads 1997, 1999; Sari et al. 1999). The clearest observational signature of a jet is an achromatic break in the light curves at t_j (e.g., Harrison et al. 1999). At radio wavelengths (i.e., below ν_m) at t_j we expect a transition from a rising $t^{1/2}$ light curve to a shallow decay of $t^{-1/3}$, while at optical wavelengths the decay is expected to steepen to t^{-p}. These decay indices refer to the asymptotic values.

Detecting a jet transition at optical wavelengths may be difficult if it occurs on timescales of a week or more. In these cases the afterglow is weak and the light from the host galaxy may start to dominate the light curve (e.g., Halpern et al. 2000). In such instances radio observations may be required to clarify matters, since the radio flux is increasing prior to t_j and changes in the light curve evolution due to the jet break are easily detected. Indeed, the jet in GRB 970508, which was very well observed in the radio is not discernible in the optical data. In this case, Frail et al. (2000c) found a wide-angle jet with an opening angle of 30° and $t_j \sim 30$ days (but see Chevalier & Li 2000).

A *ISM+Jet* model with $t_j \approx 26$ days fits the data remarkably well (see Figure 3.1). The strongest point in favor of this model is that it reproduces the broad maximum (∼1 mJy) seen from 5 GHz to 22 GHz. We expect such a plateau at t_j as all light curves for $\nu_a < \nu \leq \nu_m$ reach their peak fluxes (with only a weak $\nu^{1/3}$ frequency dependence) before undergoing a slow decline. Most other models predict a

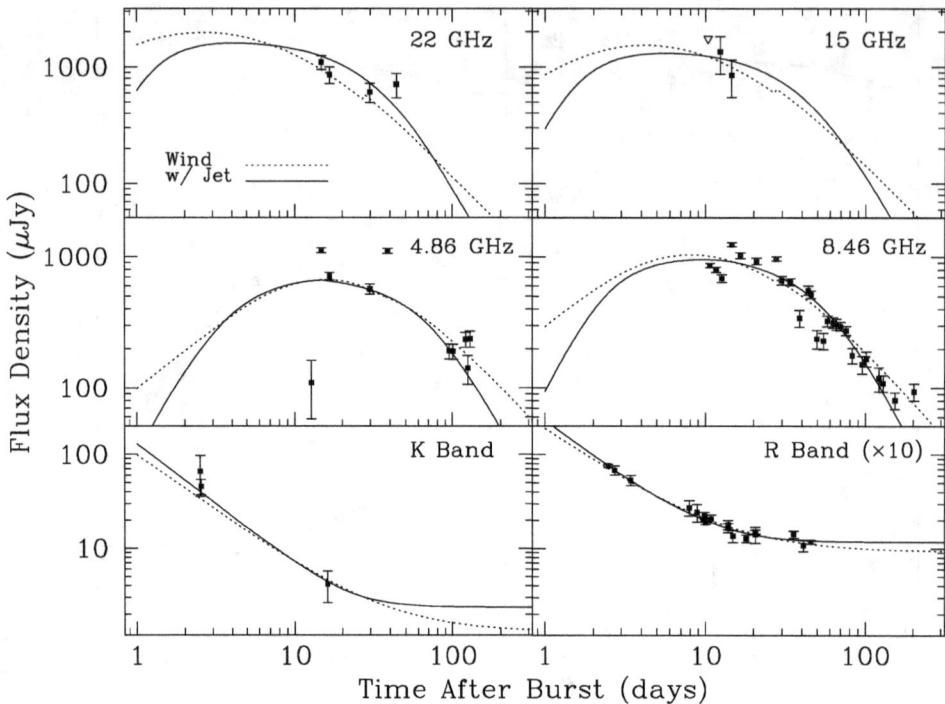

Figure 3.2: Similar to Figure 3.1 but the dotted and solid lines are light curves assuming an isotropic explosion in a wind-blown circumburst medium (Wind) and one in which the ejecta are collimated with opening angle θ_j (Wind+Jet), respectively.

strong frequency dependence in peak flux which is not seen in this case.

Knowing t_j and the density of the ambient medium n_0 from the model fit (Table 3.3) we can make a geometric correction to the total isotropic energy E_γ, as determined from either the observed γ-ray fluence or the total energy of the afterglow E_{52}, from the fit to the afterglow data. This approach gives values for the jet opening angle θ_j between $10°$ and $20°$, which for a two-sided jet reduces the GRB energy to $\sim 10^{51}$ erg. The rapid lateral expansion of the jet also accelerates the transition to the non-relativistic regime, resulting in a change in the evolution of the light curves. Since this occurs on a timescale $t_{NR} \sim t_j \theta_j^{-2} \sim 350$ days (Waxman et al. 1998), we do not expect the non-relativistic transition to be important for our data.

There is some freedom in our choice of ν_c. We know that a cooling break (i.e., $\Delta\alpha = -0.25$) is not apparent in the R band light curve on timescales of 2-10 days, so we searched for solutions with ν_c above or below this frequency. We found that physically consistent solutions (i.e., with non-negative host fluxes, and $\epsilon_B < 1$) were only possible for values of ν_c below the optical band.

As part of the fitting process we also solved for the host flux density in the R and K bands and for any local dust obscuration, assuming an LMC-like extinction law. This yields $f_{host}(R)$=1.1 μJy, $f_{host}(K)$=1.7 μJy and A_V^{host}=0.4 (in the host galaxy restframe). Klose et al. (2000b) argued for significant dust extinction with A_V^{host}=0.96. However, they likely overestimated A_V^{host} since they assumed a spherical fireball model and arbitrarily located ν_c above the optical band. Moreover, we find that there is some covariance between the values of A_V^{host} and p so that only with a global fit, in which p is constrained by the radio data as well as the optical data, we can solve for A_V^{host} in a self-consistent manner.

In view of the claims linking GRBs with the collapse of massive stars (Galama et al. 1998a; Bloom et al. 1999; Reichart 1999; Piro et al. 2000), we consider models of either spherical or jet-like explosions into a wind-blown circumburst medium (*Wind*: Chevalier & Li 1999; Li & Chevalier 1999). The wind models (Figure 3.2) fit the data as well as the *ISM+Jet* model. In fact the χ^2 is lowest for the *Wind + Jet* model. However, in view of the uncertainties in estimating the contribution of ISS to the radio flux variations (§3.4), we do not consider these differences as significant. The close match between the temporal slopes of the late-time 8.46 GHz light curve and the early R band light curve (see §3.4) is a point in favor of the *Wind* model since a steeper decline is expected for a jet geometry. Our failure to distinguish between different models of the circumburst medium can be attributed to the absence of radio measurements (particularly at millimeter wavelengths) at early times. The rapid rise of the flux density below ν_a and ν_m in the *Wind* model and the strong frequency dependence of the peak flux (see Figure 3.1), make such measurements advantageous. Moreover, in principle the *Wind* model can be distinguished from the other models by the fact that in this model ν_c is increasing with time ($\nu_c \propto t^{1/2}$). However, in this case since ν_c lies below the optical/IR bands, this behavior would be distinguishable only at late time when the host flux dominates over the OT. As before we solved for the host flux and any dust extinction (see Table 3.3).

In summary, we find that the radio and optical/NIR observations of the afterglow emission from GRB 000418 can be fit by two different models. The close similarity between the results of the *Wind* and *Jet* models has been noted for other GRBs: 970508 (Frail et al. 2000c; Chevalier & Li 2000), 980519 (Frail et al. 2000b; Chevalier & Li 1999; Jaunsen et al. 2001), 000301C (Berger et al. 2000; Li & Chevalier 2001), and 991208 (Galama et al. 2000; Li & Chevalier 2001). The resolution of this conflict is important, since it goes to the core of the GRB progenitor issue. If the GRB progenitor is a massive star then there must be evidence for a density gradient in the afterglow light curves, reflecting the stellar mass loss that occurs throughout the star's lifetime (Chevalier & Li 1999; Panaitescu & Kumar 2000). At present, an unambiguous case for a GRB afterglow expanding into a wind has yet to be found. On the contrary, most afterglows are better fit by a jet expanding in a constant density medium (e.g., Harrison et al. 1999; Halpern et al. 2000; Panaitescu & Kumar 2000) and thus we are faced with a peculiar situation. While there is good evidence linking GRBs to the dusty, gas-rich environments favored by hypernova progenitors (Bloom et al. 2002a; Galama & Wijers 2001), the expected mass loss signature is absent (or at best ambiguous) in all afterglows studied to date.

AD is supported by a Millikan Fellowship at Caltech. GRB research at Caltech is supported by NSF and NASA grants (SRK, SGD, FAH). KH is grateful for Ulysses support under JPL Contract 958056, and for NEAR support under NAG 5 9503.

Table 3.1.   Optical/Near-IR Observations of GRB 000418

UT Date	Instr.[a]	Band	Mag.[b]	Err.	Ref.[c]
Apr 20.89	TNG 3.5m	R	21.54	0.04	2
Apr 20.90	CA 3.5m	K'	17.49	0.5	2
Apr 20.93	CA 1.2m	K'	17.89	0.2	2
Apr 21.15	MDM 2.4m	R	21.66	0.12	1
Apr 21.86	LO 1.5m	R	21.92	0.14	2
Apr 26.316	USNO 1.3m	R	22.65	0.20	2,4
Apr 27.26	MDM	R	22.77	0.23	1
Apr 28.170	P200	R	22.97	0.06	1
Apr 28.3	MDM	R	22.86	0.09	1
Apr 28.413	Keck/ESI	R	23.05	0.05	1
Apr 29.26	MDM	R	22.95	0.11	1
May 2.274	Keck/ESI	Gunn-i	23.38	0.05	1
May 2.28	MDM	R	23.19	0.12	1
May 2.285	Keck/ESI	B	24.31	0.08	1
May 2.31	USNO 1.3m	R	23.11	0.130	2
May 3.26	USNO 1.3m	R	23.41	0.160	2
May 4.44	UKIRT 3.8m	K	20.49	0.40	2
May 6.42	Keck/LRIS	R	23.48	0.10	7
May 8.89	TNG	R	23.30	0.05	2
May 8.92	TNG	V	23.92	0.07	2
May 9.82	USNO 1.0m	R	23.37	0.21	2
May 23.93	TNG	R	23.37	0.10	2
May 29.228	P200	R	23.66	0.15	1
Jun 2.88	CA 3.5m	R	23.32	0.08	2
Jun 2.91	TNG	R	23.57	0.05	2

[a]CA 3.5m=Calar Alto 3.5-meter, USNO1.3m=U.S. Naval Observatory Flagstaff Station 1.3-meter, ESI=W.M. Keck Observatory Echellette Spectrograph-Imager, LRIS=W.M. Keck Observatory Low-Resolution Imaging Spectrograph

[b]Optical photometry is on the Kron-Cousins and Gunn systems and referred to that of Henden (2000). Data are corrected for Galactic extinction corresponding to $E(B - V) = 0.032$ derived from the maps of Schlegel et al. (1998).

[c]1=this work, 2=Klose et al. (2000b), 3=Henden et al. (2000), 4=Metzger & Fruchter (2000)

Table 3.2. Radio Observations of GRB 000418

Epoch (UT)	Telescope	ν_0 (GHz)	$S \pm \sigma$ (μJy)	Epoch (UT)	Telescope	ν_0 (GHz)	$S \pm \sigma$ (μJy)
2000 Apr 28.75	Ryle	15.0	550±600	2000 Jun 3.04	VLA	8.46	517±34
2000 Apr 29.07	VLA	8.46	856±33	2000 Jun 7.01	VLA	8.46	238±38
2000 Apr 30.07	VLA	8.46	795±37	2000 Jun 11.93	VLA	8.46	230±33
2000 Apr 30.73	Ryle	15.0	1350±480	2000 Jun 15.13	VLA	8.46	325±30
2000 May 1.06	VLA	4.86	110±52	2000 Jun 20.10	VLA	8.46	316±30
2000 May 1.06	VLA	8.46	684±48	2000 Jun 23.19	VLA	8.46	306±29
2000 May 2.93	Ryle	15.0	850±300	2000 Jun 27.08	VLA	8.46	296±22
2000 May 3.04	VLA	4.86	1120±52	2000 Jul 2.98	VLA	8.46	274±22
2000 May 3.04	VLA	8.46	1240±46	2000 Jul 10.04	VLA	8.46	178±24
2000 May 3.04	VLA	22.46	1100±150	2000 Jul 22.81	VLA	8.46	152±23
2000 May 4.97	VLA	1.43	210±180	2000 Jul 22.81	VLA	4.86	192±25
2000 May 4.97	VLA	4.86	710±47	2000 Jul 28.50	VLA	8.46	168±22
2000 May 4.97	VLA	8.46	1020±53	2000 Jul 28.50	VLA	4.86	191±25
2000 May 4.97	VLA	22.46	860±141	2000 Aug 17.74	VLA	8.46	119±25
2000 May 7.18	VLBA	8.35	625±60	2000 Aug 17.74	VLA	4.86	235±31
2000 May 9.25	VLA	8.46	926±53	2000 Aug 21.65	VLA	4.86	142±35
2000 May 16.13	VLA	8.46	963±34	2000 Aug 21.65	VLA	8.46	87±31
2000 May 18.24	VLA	4.86	567±50	2000 Aug 25.78	VLA	4.86	238±34
2000 May 18.24	VLA	8.46	660±50	2000 Aug 25.78	VLA	8.46	166±27
2000 May 18.24	VLA	22.46	610±114	2000 Aug 27.89	VLA	8.46	100±25
2000 May 22.21	VLA	8.46	643±38	2000 Sep 10.73	VLA	8.46	148±25
2000 May 26.92	VLA	4.86	1105±51	2000 Sep 18.68	VLA	8.46	55±20
2000 May 26.92	VLA	8.46	341±50	2000 Sep 26.62	VLA	8.46	85±22
2000 Jun 1.14	VLA	8.46	556±43	2000 Nov 6.55	VLA	8.46	94±14
2000 Jun 1.14	VLA	22.46	710±16				

Note. — The columns are (left to right), (1) UT date of the start of each observation, (2) telescope name, (3) observing frequency, and (4) peak flux density at the best fit position of the radio transient, with the error given as the root mean square noise on the image.

Table 3.3.   Synchrotron Model Parameters for GRB 000418

Parameters[a]	ISM	ISM+Jet	Wind	Wind+Jet
ν_a (Hz)	4.1×10^9	1.7×10^9	30×10^9	3.7×10^9
ν_m (Hz)	2.3×10^{11}	1.8×10^{10}	5.8×10^{11}	1.1×10^{11}
ν_c (Hz)	2×10^{15}	10^{14}	1.8×10^{13}	5×10^{12}
f_m (mJy)	2.5	3.4	10.4	3.7
p	2.3	2.4	2.2	2.5
t_j (days)	$\cdots$	25.7	$\cdots$	14.6
A_V^{host}	0.0	0.4	0.3	0.2
χ^2/dof	326/54	165/53	184/53	127/53
E_{52}	11	10	4	1.6
n_0 or A^*	0.01	0.02	0.14	0.07
ϵ_B	0.05	0.06	0.04	0.70
ϵ_e	0.03	0.10	0.07	0.14

[a]For the *ISM* and *Wind* models ν_a, ν_m, ν_c and f_m are the self-absorption, synchrotron peak, and cooling frequencies, and the peak flux density, respectively on day 1. For the *ISM+Jet* and *Wind+Jet* model these values are referenced instead to the jet break time t_j. p is the electron power-law index and A_V is the V band extinction in the rest frame of the host galaxy (z=1.118), assuming an LMC-like extinction curve. The resulting values of χ^2 include an estimated contribution of interstellar scattering (ISS) and the increased error in subtracting off a host galaxy flux from each of the optical points. The model parameters are the total isotropic energy E_{52} in units of 10^{52} erg, the ambient density n_0 in cm^{-3} or in the case of the two wind models the parameter A^* as defined by Chevalier & Li (1999). ϵ_e and ϵ_B are the fraction of the shock energy in the electrons and the magnetic field, respectively. The true uncertainties in the derived parameters are difficult to quantify due to covariance, but we estimate that they range from $10-20\%$

CHAPTER 4

A Standard Kinetic Energy Reservoir in Gamma-Ray Burst Afterglows[†]

E. BERGER[a], S. R. KULKARNI[a], & D. A. FRAIL[b]

[a]Department of Astronomy, 105-24 California Institute of Technology, Pasadena, CA 91125, USA

[b]National Radio Astronomy Observatory, P. O. Box 0, Socorro, NM 87801

Abstract

We present a comprehensive sample of X-ray observations of 41 γ-ray burst (GRB) afterglows, as well as jet opening angles, θ_j for a subset with measured jet breaks. We show that there is a significant dispersion in the X-ray fluxes, and hence isotropic X-ray luminosities ($L_{X,\mathrm{iso}}$), normalized to $t = 10$ hr. However, there is a strong correlation between $L_{X,\mathrm{iso}}$ and the beaming fractions, $f_b \equiv [1 - \cos(\theta_j)]$. As a result, the true X-ray luminosity of GRB afterglows, $L_X = f_b L_{X,\mathrm{iso}}$, is approximately constant, with a dispersion of only a factor of two. Since $\epsilon_e E_b \propto L_X$, the strong clustering of L_X directly implies that the adiabatic blastwave kinetic energy in the afterglow phase, E_b, is tightly clustered. The narrow distribution of L_X also suggests that $p \approx 2$, that inverse Compton emission does not in general dominate the observed X-ray luminosity, and that radiative losses at $t < 10$ hr are relatively small. Thus, despite the large diversity in the observed properties of GRBs and their afterglows the energy imparted by the GRB central engine to the relativistic ejecta is approximately constant.

SECTION 4.1

Introduction

Gamma-ray bursts (GRBs) exhibit a remarkable diversity: Fluences range from 10^{-7} to 10^{-3} erg cm^{-2}, peak energies range from 50 keV to an MeV, and possibly from the X-ray to the GeV band (Fishman & Meegan 1995), and durations extend from about 2 to 10^3 s (for the long-duration GRBs). This diversity presumably reflects a dispersion in the progenitors and the properties of the central engine. Perhaps the most impressive feature of GRBs are their brilliant luminosities and isotropic energy releases approaching the rest mass of a neutron star, $E_{\gamma,\mathrm{iso}} \sim 10^{54}$ erg (Kulkarni et al. 1999a; Andersen et al. 2000).

The quantity of energy imparted to the relativistic ejecta, E_{rel}, and the quality parameterized by the bulk Lorentz factor, Γ, are the two fundamental properties of GRB explosions. In particular, extremely

---

[†] A version of this chapter was published in *The Astrophysical Journal*, vol. 590, 379–385, (2003).

high energies push the boundaries of current progenitor and engine models, while low energies could point to a population of sources that is intermediate between GRBs and core-collapse supernovae.

The true energy release depends sensitively on the geometry of the ejecta. If GRB explosions are conical (as opposed to spherical) then the true energy release is significantly below that inferred by assuming isotropy. Starting with GRB 970508 (Waxman et al. 1998; Rhoads 1999) there has been growing observational evidence for collimated outflows, coming mainly from achromatic breaks in the afterglow light curves.

In the conventional interpretation, the epoch at which the afterglow light curves steepen ("break") corresponds to the time at which Γ decreases below θ_j^{-1}, the inverse opening angle of the collimated outflow or "jet" (Rhoads 1999). The break happens for two reasons: an edge effect, and lateral spreading of the jet which results in a significant increase of the swept up mass. Many afterglows have $t_j \sim 1-$ few days, which are best measured from optical/near-IR light curves (e.g., Harrison et al. 1999; Kulkarni et al. 1999a; Stanek et al. 1999), while wider opening angles are easily measured from radio light curves (e.g., Waxman et al. 1998; Berger et al. 2001a).

Recently, Frail et al. (2001) inferred θ_j for fifteen GRB afterglows from measurements of t_j and found the surprising result that $E_{\gamma,\mathrm{iso}}$ is strongly correlated with the beaming factor, f_b^{-1}; here, $f_b \equiv [1 - \cos(\theta_j)]$ is the beaming fraction and $E_{\gamma,\mathrm{iso}}$ is the γ-ray energy release inferred by assuming isotropy. In effect, the true γ-ray energy release, $E_\gamma = f_b E_{\gamma,\mathrm{iso}}$ is approximately the same for all the GRBs in their sample, with a value of about 5×10^{50} erg (assuming a constant circumburst density, $n_0 = 0.1$ cm^{-3}). In the same vein, broad-band modeling of several GRB afterglows indicates that the typical blastwave kinetic energy in the adiabatic afterglow phase is $E_b \sim 5 \times 10^{50}$ erg, with a spread of about 1.5 orders of magnitude (Panaitescu & Kumar 2002). However, the general lack of high quality afterglow data severely limits the application of the broad-band modeling method.

Separately, Kumar (2000) and Freedman & Waxman (2001) noted that the afterglow flux at frequencies above the synchrotron cooling frequency, ν_c, is proportional to $\epsilon_e dE_b/d\Omega$, where ϵ_e is the fraction of the shock energy carried by electrons and $dE_b/d\Omega$ is the energy of the blastwave per unit solid angle. The principal attraction is that the flux above ν_c does not depend on the circumburst density, and depends only weakly on the fraction of shock energy in magnetic fields, ϵ_B. For reasonable conditions (which have been verified by broad-band afterglow modeling, e.g., Panaitescu & Kumar 2002), the X-ray band ($2 - 10$ keV) lies above ν_c starting a few hours after the burst. Thus, this technique offers a significant observational advantage, namely the X-ray luminosity can be used as a surrogate for the isotropic-equivalent afterglow kinetic energy.

Piran et al. (2001) find that the X-ray flux, estimated at a common epoch ($t = 11\,\mathrm{hr}$), exhibits a narrow distribution of $\log(F_X)$, $\sigma_l(F_X) = 0.43^{+0.12}_{-0.11}$; here $\sigma_l^2(x)$ is the variance of $\log(x)$. Taken at face value, the narrow distribution of F_X implies a narrow distribution of $\epsilon_e dE_b/d\Omega$. This result, if true, is quite surprising since if the result of Frail et al. (2001) is accepted then $dE_b/d\Omega$ should show a wide dispersion comparable to that of f_b^{-1}.

Still, Piran et al. (2001) extend their statistical analysis with the following argument. The relation between $dE_b/d\Omega$ and E_b can be restated as $\log(dE_b/d\Omega) = \log(E_b) + \log(f_b^{-1})$. Thus, $\sigma_l^2(dE_b/d\Omega) = \sigma_l^2(E_b) + \sigma_l^2(f_b^{-1})$. Since $dE_b/d\Omega \propto L_{X,\mathrm{iso}}$ (for a constant ϵ_e) they express, $\sigma_l^2(E_b) = \sigma_l^2(L_{X,\mathrm{iso}}) - \sigma_l^2(f_b^{-1})$. Given the diversity in θ_j (Frail et al. 2001) and the apparent narrowness in F_X (above), it would then follow that E_b should be very tightly clustered.

However, the approach of Piran et al. (2001) makes a key assumption, namely that E_b and f_b^{-1} are uncorrelated. This is certainly true when E_b is constant, but the assumption then pre-supposes the answer! In reality, a correlation between E_b and f_b can either increase or decrease $\sigma_l^2(E_b)$, and this must be addressed directly. In addition, as appears to be the case (see §4.2), $\sigma_l^2(f_b^{-1})$ is dominated by bursts with the smallest opening angles, which results in a distinctly different value than the one used by Piran et al. (2001) based only on the observed values of θ_j.

In this paper, we avoid these concerns by taking a direct approach: we measure the variance in $E_b \propto f_b L_{X,\mathrm{iso}}$ rather than bounding it through a statistical relation. We show, with a larger sample, that $L_{X,\text{iso}}$ is not as narrowly distributed as claimed by Piran et al. (2001), and in fact shows a spread similar to that of $E_{\gamma,\text{iso}}$. On the other hand, we find that $L_{X,\text{iso}}$ is strongly correlated with f_b^{-1}. It is this correlation, and not the claimed clustering of $L_{X,\text{iso}}$, that results in, and provides a physical basis for the strong clustering of L_X and hence the blastwave kinetic energy, E_b.

SECTION 4.2

X-ray Data

In Table 4.1 we provide a comprehensive list of X-ray observations for 41 GRB afterglows, as well as temporal decay indices, α_X ($F_\nu \propto t^{\alpha_X}$), when available. In addition, for a subset of the afterglows for which jet breaks have been measured from the radio, optical, and/or X-ray emission, we also include the inferred θ_j (Frail et al. 2001; Panaitescu & Kumar 2002). We calculate θ_j from t_j using the circumburst densities inferred from broad-band modeling, when available, or a fiducial value of $10\,\text{cm}^{-3}$, as indicated by the best-studied afterglows (e.g., Yost et al. 2002). This normalization for n_0 is different from Frail et al. (2001) who used $n_0 = 0.1\,\text{cm}^{-3}$.

For all but one burst we interpolate the measured F_X to a fiducial epoch of 10 hr (hereafter, $F_{X,10}$), using the measured α_X when available, and the median of the distribution, $\langle\alpha_X\rangle = -1.33 \pm 0.38$ when a measurement is not available. The single exception is GRB 020405 for which the first measurement was obtained $t \approx 41$ hr, while the inferred jet break time is about 23 hr (Berger et al. 2003d). In this case, we extrapolate to $t = 10$ hr using $\alpha_X = -1.69$ for $t > 23$ hr and $\alpha_X = -0.78$ for $t < 23$ hr. We list the values of $F_{X,10}$ in Table 4.2.

In Figure 4.1 we plot the resulting distribution of $F_{X,10}$. For comparison we also show the distribution of γ-ray fluences from the sample presented by Bloom et al. (2001) and updated from the literature. Clearly, while the distribution of X-ray fluxes is narrower than that of the γ-ray fluences, $\sigma_l(f_\gamma) = 0.79^{+0.10}_{-0.08}$, it still spans ~ 2.5 orders of magnitude, i.e., $\sigma_l(F_{X,10}) = 0.57^{+0.07}_{-0.06}$. The value of $\sigma_l(F_{X,10})$, and all variances quoted below, are calculated by summing the Gaussian distribution for each measurement, and then fitting the combined distribution with a Gaussian profile.

We translate the observed X-ray fluxes to isotropic luminosities using:

$$L_{X,\text{iso}}(t = 10\,\text{hr}) = 4\pi d_L^2 F_{X,10}(1 + z)^{\alpha_X - \beta_X - 1}. \tag{4.1}$$

We use $\beta_X \approx -1.05$, the weighted mean value for X-ray afterglows (De Pasquale et al. 2002), and the median redshift, $\langle z \rangle = 1.1$, for bursts that do not have a measured redshift. The resulting distribution of $L_{X,\text{iso}}$, $\sigma_l(L_{X,\text{iso}}) = 0.68^{+0.17}_{-0.09}$, is wider than that of F_X due to the dispersion in redshifts. We note that this is wider than the value quoted by Piran et al. (2001) of $\sigma_l(L_{X,\text{iso}}) \approx 0.43$ based on a smaller sample, and ignoring the dispersion in redshift. Using the same method we find $\sigma_l(E_{\gamma,\text{iso}}) = 0.92^{+0.12}_{-0.08}$.

In the absence of a strong correlation between f_b and $L_{X,\text{iso}}$, the above results indicate that the distribution of the true X-ray luminosities, $L_X \equiv f_b^{-1} L_{X,\text{iso}}$, should have a wider dispersion than either $L_{X,\text{iso}}$ or f_b, for which we find $\sigma_l(f_b) = 0.52^{+0.13}_{-0.12}$ (Frail et al. 2001). Instead, when we apply the individual beaming corrections for those bursts that have a measured θ_j and redshift[1] (see Table 4.2), we find a significantly narrower distribution, $\sigma_l(L_X) = 0.32^{+0.10}_{-0.06}$.

SECTION 4.3

Beaming Corrections and Kinetic Energies

The reduced variance of L_X compared to that of $L_{X,\text{iso}}$ requires a strong correlation between $L_{X,\text{iso}}$ and f_b^{-1}, such that bursts with a brighter isotropic X-ray luminosity are also more strongly collimated. Indeed, as can be seen from Figure 4.2 the data exhibit such a correlation. Ignoring the two bursts which are obvious outliers (980326 and 990705), as well as GRBs 980329 and 980519, which do not

---

[1] These do not include GRB 990705 which is poorly characterized; see §4.3.

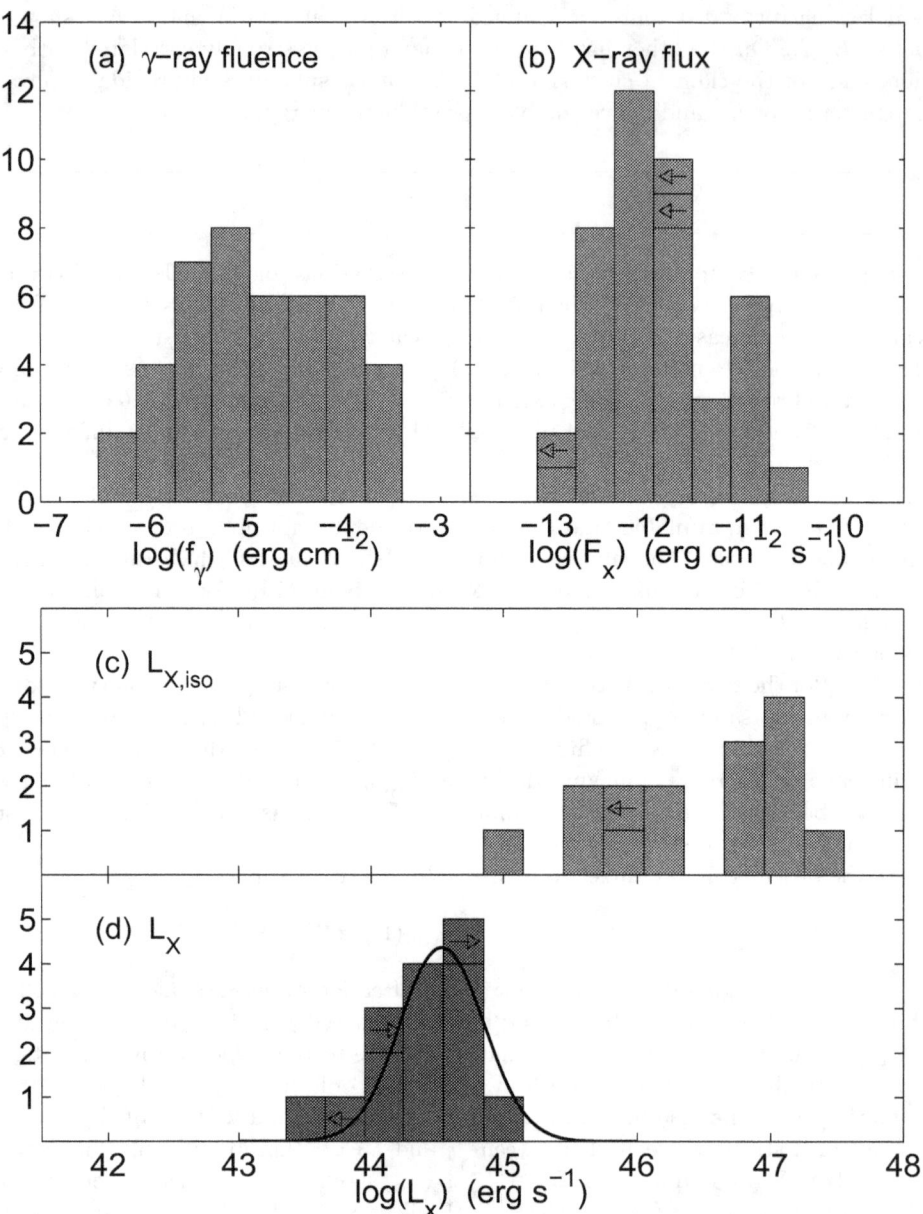

Figure 4.1: Panel (a) shows the distribution of γ-ray fluences. Panel (b) shows the distribution of X-ray fluxes scaled to $t = 10$ hr after the burst. In panel (c) we plot the isotropic-equivalent X-ray luminosity, $L_{X,\mathrm{iso}}$, for the subset of X-ray afterglows with known θ_j and redshift, while in panel (d) we show the true X-ray luminosity, $L_X = f_b^{-1} L_{X,\mathrm{iso}}$.

have a measured redshift, we find $L_{X,\mathrm{iso}} \propto f_b^{-0.80}$. The linear correlation coefficient between $\log(L_{X,\mathrm{iso}})$ and $\log(f_b^{-1})$ indicates a probability that the two quantities are not correlated of only 4.6×10^{-4}. For $\log(E_{\gamma,\mathrm{iso}})$ and $\log(f_b^{-1})$ we find a similar probability of 4.2×10^{-4} that the two quantities are not correlated.

Thus, as with the γ-ray emission, the observed afterglow emission also exhibits strong luminosity diversity due to strong variations in f_b. Therefore, the mystery of GRBs is no longer the energy release but understanding what aspect of the central engine drives the wide diversity of f_b.

We note that there are four possible outliers in the correlation between $L_{X,\mathrm{iso}}$ and f_b^{-1}. The afterglows of GRBs 980326 and 980519 exhibit rapid fading (Groot et al. 1998; Vrba et al. 2000), which has been interpreted as the signature of an early jet break. However, it is possible that the rapid fading is instead due to a $\rho \propto r^{-2}$ density profile, and in fact for GRB 980519 such a model indicates $\theta_j \approx 0.12$, three times wider than in the constant density model. This is sufficient to bring GRB 980519 into agreement with the observed correlation. The redshift of GRB 980329 is not known, but with $z = 2$ it easily agrees with the correlation. Finally, the X-ray flux and jet opening angle for GRB 990705 are poorly characterized due to contamination from a nearby source (De Pasquale et al. 2002) and a poor optical light curve (Masetti et al. 2000b).

SECTION 4.4

Discussion and Conclusions

We have presented a comprehensive compilation of early X-ray observations of 41 GRBs, from which we infer $F_{X,10}$, the flux in the 2–10 keV band at 10 hr. As first pointed by Kumar (2000) and Freedman & Waxman (2001), the afterglow luminosity above the cooling frequency is $L_{X,iso} \propto \epsilon_e E_{b,\mathrm{iso}}$ where $E_{b,\mathrm{iso}}$ is the isotropic-equivalent explosion kinetic energy. More importantly, the flux is independent of the ambient density and weakly dependent on ϵ_B. For all well-modeled afterglows, the cooling frequency at 10 hr is below the X-ray band. Thus, $F_{X,10}$ can be utilized to yield information about the kinetic energy of GRBs.

Earlier work (Piran et al. 2001) focussed on statistical studies of $F_{X,10}$ and found the very surprising result that it is narrowly clustered. By assuming that the true kinetic energy, $E_b = E_{b,\mathrm{iso}}f_b \propto L_X = L_{X,\mathrm{iso}}f_b$, and f_b (the beaming factor) are uncorrelated, the authors deduced that L_X and thus E_b are even more strongly clustered. However, this approach is weakened by assuming (in effect) the answer. Furthermore, the approach of Piran et al. (2001) which relies on subtracting variances is very sensitive to measurement errors. To illustrate this point, we note $\sigma_l^2(L_{X,\mathrm{iso}}) = 0.68^{+0.17}_{-0.09}$ for the entire sample presented here, whereas $\sigma_l^2(f_b) = 0.52^{+0.13}_{-0.12}$. Thus, $\sigma_l^2(L_X) = 0.16^{+0.30}_{-0.21}$ may be negative using the statistical approach.

In contrast to the statistical approach, we take the direct approach and estimate the true kinetic energy, $E_b \propto L_{X,\mathrm{iso}}f_b$, by using the measured $L_{X,\mathrm{iso}}$ and inferred f_b. The advantage of our approach is that we do not make assumptions of correlations (or lack thereof) and more importantly we do not subtract variances. We directly compute the variance of the desired physical quantity, namely L_X, and find that it is strongly clustered.

Even more importantly, with our direct approach we have uncovered the physical reason for the wide dispersion in $L_{X,\mathrm{iso}}$ and the clustering of L_X, namely the dispersion in jet opening angles.

L_X is related to the physical quantities as follows (Freedman & Waxman 2001):

$$\epsilon_e E_b \propto A L_X Y^\epsilon, \tag{4.2}$$

where

$$Y \equiv B \epsilon_e^{-3} \epsilon_B^{-1} L_{X,\mathrm{iso}}^{-1}. \tag{4.3}$$

Here $\epsilon \equiv (p-2)/(p-1)$, as well as A and B depend to some extent on the details of the electron distribution (power law versus relativistic Maxwellian; the value of power law index, p).

There is no reason to expect that L_X should be clustered. However, one can argue that the microphysics should be the same for each GRB afterglow, in particular ϵ_e and p. The best studied afterglows appear to favor $p = 2.2$ (e.g., Frail et al. 2000c; Galama et al. 1998d), a value also favored by our current theoretical knowledge of shock acceleration (see Ostrowski & Bednarz 2002 and references therein). In addition, as already indicated by the γ-ray observations, there is evidence supporting strong clustering

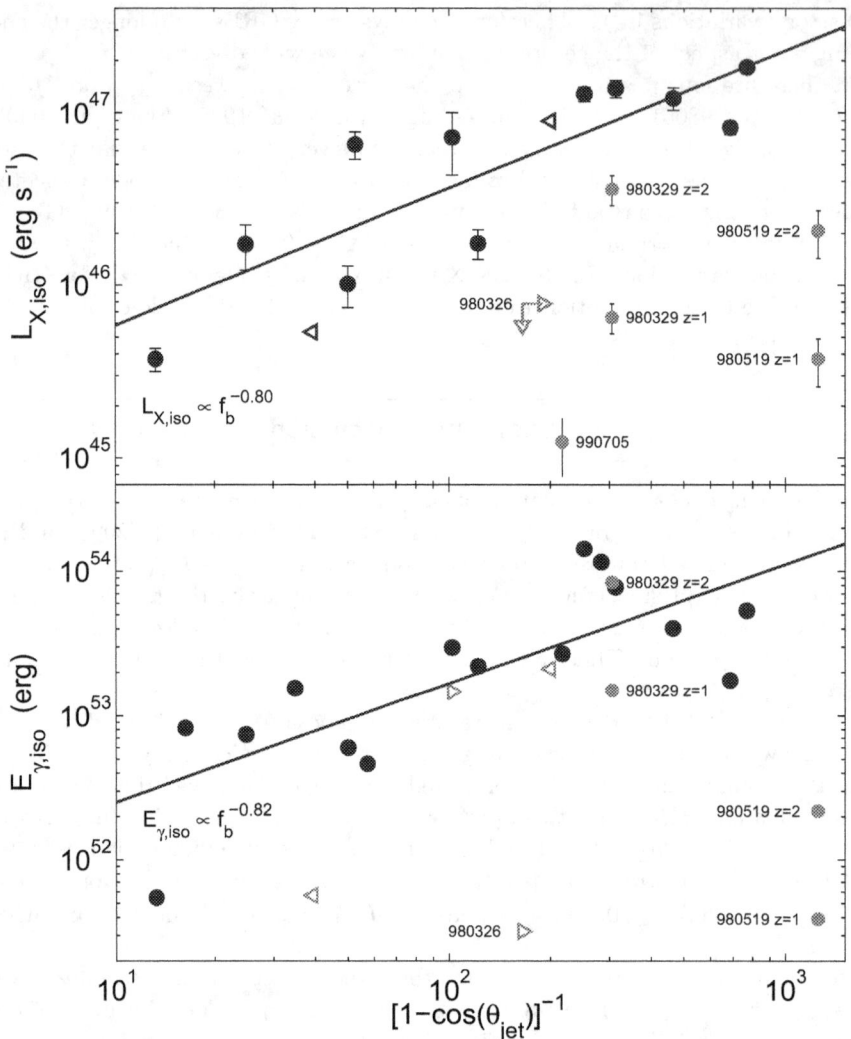

Figure 4.2: Isotropic-equivalent X-ray luminosity (top) and isotropic-equivalent γ-ray energy (bottom) as a function of the beaming factor, $[1 - \cos(\theta_j)]^{-1}$. There is a strong positive correlation between $L_{X,\mathrm{iso}}$ and f_b^{-1}, as well as between $E_{\gamma,\mathrm{iso}}$ and f_b^{-1} resulting in an approximately constant true X-ray luminosity and γ-ray energy release. In fact, while the distributions of all three parameters span about three orders of magnitude, the distributions of the beaming-corrected parameters span about one order of magnitude.

of explosion energies in GRBs (Frail et al. 2001).

    Given these reasonable assumptions, a strong clustering of L_X makes sense if the physical quantities that are responsible for L_X are clustered. As can be seen from Equation 4.2, this would require that L_X be linearly related to E_b. Such a relation is possible if four conditions are met.

    First, the afterglow X-ray emission on timescales of 10 hr must be primarily dominated by synchrotron emission (which is the basis of Equation 4.2). Contribution from inverse Compton (IC) emission, which depends strongly on n_0 and ϵ_B (Sari & Esin 2001), is apparently not significant. A possible exception is GRB 000926 (Harrison et al. 2001), but even there the IC contribution is similar to that from synchrotron emission.

Second, the energy radiated by the afterglow from the time of the explosion to $t = 10$ hr cannot be significant. This constrains the radiative losses at early time to at most a factor of few.

Third, p must be relatively constant (as one may expect in any case from insisting that the microphysics should not be different for different bursts). For example, changing p from a value of 1.5 to 3 results in Y^ϵ ranging from 0.003 to 117, a factor of 39,000! Even small changes in p, e.g., from $p = 1.75$ to $p = 2.25$, result in a factor of 8 change in Y^ϵ. In contrast, some afterglow models yield values of p significantly below 2 (e.g., Panaitescu & Kumar 2002), while others have p approaching 3 (Chevalier & Li 2000). Our results, on the other hand, indicate that one should set $p \approx 2$ and attribute apparent deviant values of p to external environment or energy injection from the central source.

Finally, since both the prompt and afterglow emission exhibit a strong correlation with f_b, which is determined from late-time observations (hours to weeks after the burst), the resulting constancy of both E_γ and E_b, indicates that GRB jets must be relatively homogeneous and maintain a simple conical geometry all the way from internal shocks ($\sim 10^{13} - 10^{14}$ cm) to the epoch of jet break ($\sim 10^{17}$ cm). This rules out the idea that brighter bursts are due to bright spots along specific lines of sight (Kumar & Piran 2000). At the same time, the possible deviation from a linear relation between $\log(L_{X,\text{iso}})$ and $\log(f_b^{-1})$ may hold a clue to the structure of the jet.

With the result that GRB afterglows have a standard kinetic energy firmly established, the next step is to closely investigate bursts that deviate from this relation; such sources may be a clue to sub-classes of GRBs (e.g., Bloom et al. 2003b). Fortunately, while the statistical study of afterglow energetics used previously misses this point completely, the direct method employed in this paper can easily uncover these sources. More importantly, this method provides a framework for understanding the underlying physical processes which may give rise to such a diversity.

SRK thanks S. Phinney for valuable discussions. We also thank D. Lazzati, B. Zhang, and the anonymous referee for valuable comments. We acknowledge support from NSF and NASA grants.

Table 4.1. X-ray Afterglow Data

GRB	z	Epoch (hrs)	Flux (10^{-13} erg/cm^2/s)	α_X	θ_{jet}	Ref.
970111	...	24.0	1.05±0.46	−0.4±3.2[a]	...	1,2
		30.7	0.95±0.34	...	...	2
970228	0.695	8.5	33.8±3.3	−1.27±0.14	...	2,3
		12.7	28±4			2
		92.4	1.5±0.4			2
970402	...	9.9	2.9±0.4	−1.35±0.55	...	2
		16.8	1.5±0.4			2
970508	0.835	13.1	7.13	−1.1±0.1	0.391	4,5
		72.3	4.3±0.5			2
		104	2.3±0.7			2
970815	...	89.6	<1	...	...	6
970828	0.958	4.0	118	−1.42	0.128	5,7
		42.6	4.1			7
971214	3.418	8.1	9.0±0.9	−1±0.2	>0.100	2,5
		28.9	2.1±0.4			2
971227	...	16.5	2.5±0.7	−1.12±0.06	...	8
980326	~1[b]	8.5	<16	...	<0.110	9
980329	...	8.4	14±2.1	−1.55±0.3	0.081	10,11
		11.8	6.2±1.2			10
		16.4	3.4±1.0			10
		23.7	2.7±0.7			10
		43.6	1.1±0.4			10
980515	...	11	$2.0^{+0.5}_{-0.9}$	...	...	12
980519	<2[c]	10.9	5.3±1.0	−1.7±0.7	0.040	13,14
		15.3	2.0±0.4			13
		21.5	1.6±0.5			13
		27.2	0.8±0.4			13
980613	1.096	9.9	7.1±1.9	−0.92±0.62	>0.226	2
		23.4	4.0±0.8			2
980703	0.966	34.0	4.0±1	−1.24±0.18	0.200	2,15
981226	...	14.0	4.0	−1.3±0.4	...	16
990123	1.600	6.4	124±11	−1.41±0.05	0.089	2,5
		23.4	19.1±2.2			2
990217	...	11	<1.1	...	...	12
990510	1.619	8.7	47.8±3.1	−1.41±0.18	0.054	5,14,17
		10.1	40.5±2.6			17
990510		17.1	18.5±3.1			17
		19.1	20.9±2.3			17
		24.0	12.1±1.4			17
		26.3	9.9±1.1			17
		29.4	7.8±1.1			17
990627	...	11.9	3.5	...	...	18
990704	...	10.1	10.1±2.9	−1.3±0.3	...	19
		13.4	8.9±2.2			19
		23.3	3.1±2.0			19
		26.8	2.9±1.6			19
990705	0.840	14.5	1.9±0.6	...	0.096	5,20
990806	...	13.6	5.5±1.5	−1.4±0.7	...	21
		34.3	1.5±0.6			21
990907	...	11	10.2±5.6	...	...	12
991014	...	11	$4.0^{+1.4}_{-1.2}$	...	...	12
991216	1.020	4.0	1240±40	−1.61±0.07	0.051	5,14,22
		10.9	250±10			22
000115	...	2.9	270	<−1	...	23
000210	0.846	11	4.0±1.0	−1.38±0.03	...	24
000214	...	14.9	5	−1.8	...	25
		22.1	2.5			25
000528	...	11	2.3±1.0	...	...	12
000529	...	9.0	2.8	...	...	26
000926	2.037	54.9	2.23±0.77	−3.7±1.5[a]	0.140	14,27
		66.5	0.94±0.14			27
001025	...	50.4	0.53±0.10	−3±1.9[a]	...	28
001109	...	19.3	7.1±0.5	...	...	29
010214	...	7.7	6	<−1.6	...	30
		24.1	<0.5			30
010220	...	20.8	0.33	−1.2±1.0	...	28
010222	1.477	8.9	101±11	−1.33±0.04	0.080	14,31
		32.7	18.7±1.8			31
		54.4	9.9±0.5			31
011211	2.14	11.0	1.9	−1.7±0.2	...	32
020322	...	18.8	3.5±0.2	−1.26±0.23	...	33
020405	0.698	41.0	13.6±2.5	−1.15±0.95[d]	0.285	34,35,36

Table 4.1

GRB	z	Epoch (hrs)	Flux (10^{-13} erg/cm^2/s)	α_X	θ_{jet}	Ref.	GRB	z	Epoch (hrs)	Flux (10^{-13} erg/cm^2/s)	α_X	θ_{jet}	Ref.
		11.7	32.8 ± 3.7			17	020813	1.254	31.9	22	-1.42 ± 0.05	0.066	37,38
		13.4	22.8 ± 2.8			17	021004	2.323	31.4	4.3 ± 0.7	-1.0 ± 0.2	0.240	39,40
		15.3	24.1 ± 2.7			17							

Note. — The columns are (left to right): (1) GRB name, (2) redshift, (3) mid-point epoch of X-ray observation, (4) X-ray flux, (5) temporal decay index ($F_X \propto t^{\alpha_X}$), (6) jet opening angle, and (7) references for the X-ray flux and jet opening angle. ^{a} Due to the large uncertainty in the value of α_X we use the median value for the sample, $\langle \alpha_X \rangle = -1.33 \pm 0.38$. ^{b} The redshift is based on matching the optical light curve of SN 1998bw to the red excess reported by Bloom et al. (1999). ^{c} The redshift limit is based on a detection of the afterglow in the optical U-band (Jaunsen et al. 2001). ^{d} The inferred jet break is at $t = 0.95$, prior to the X-ray observation — we use the model fit to extrapolate the flux to $t = 10$ hr (Berger et al. in prep.).

References. — (1) Feroci et al. (1998); (2) Piro (2001); (3) Frontera et al. (1998); (4) Piro et al. (1998); (5) Frail et al. (2001); (6) Murakami et al. (1997); (7) Smith et al. (2002a); (8) Antonelli et al. (1999); (9) Marshall & Takeshima (1998); (10) in 't Zand et al. (1998); (11) Yost et al. (2002); (12) De Pasquale et al. (2002); (13) Nicastro et al. (1999a); (14) Panaitescu & Kumar (2002); (15) Vreeswijk et al. (1999); (16) Frontera et al. (2000); (17) Pian et al. (2001); (18) Nicastro et al. (1999b); (19) Feroci et al. (2001); (20) Amati et al. (2000b); (21) Frontera et al. (1999); (22) Takeshima et al. (1999); (23) Marshall et al. (2000); (24) Piro et al. (2002); (25) Antonelli et al. (2000); (26) Feroci et al. (2000); (27) Harrison et al. (2001); (28) Watson et al. (2002); (29) Amati et al. (2000a); (30) Frontera et al. (2001); (31) in 't Zand et al. (2001); (32) Reeves et al. (2002); (33) Watson et al. (2002); (34) Price & et al. (2002); (35) Mirabal et al. (2002); (36) Berger et al. (in prep); (37) Price et al. (2002a); (38) Vanderspek et al. (2002); (39) Fox et al. (in prep); (40) Frail et al. (in prep).

Table 4.2.   X-ray Afterglow Data at $t = 10$ hr

GRB	z	$F_{X,10}$ (10^{-13} erg/cm^2/s)	$L_{X,iso}$ (10^{45} erg s^{-1})	θ_{jet}	L_X (10^{44} erg s^{-1})
970111	...	3.36 ± 1.64	2.56 ± 1.25	...	...
970228	0.695	27.50 ± 3.17	6.82 ± 0.79	...	...
970402	...	2.86 ± 0.61	2.18 ± 0.46	...	...
970508	0.835	9.60 ± 1.47	3.74 ± 0.57	0.391	2.82 ± 0.43
970815	...	< 18.47	< 14.1	...	...
970828	0.958	32.12 ± 6.31	17.6 ± 3.4	0.128	1.44 ± 0.28
971214	3.418	7.29 ± 0.87	89.6 ± 10.8	> 0.100	> 4.48
971227	...	4.38 ± 1.26	3.34 ± 0.96	< 0.110	< 0.59
980326	~1	< 12.89	< 9.82	...	...
980329	...	10.68 ± 2.10	8.14 ± 1.60	0.081	0.27 ± 0.05
980515	...	2.27 ± 0.90	1.73 ± 0.69	...	...
980519	...	6.14 ± 1.89	4.68 ± 1.44	0.040	0.04 ± 0.01
980613	1.096	7.03 ± 2.28	5.36 ± 1.74	> 0.226	> 1.36
980703	0.966	18.24 ± 4.97	10.2 ± 2.8	0.200	2.03 ± 0.55
981226	...	6.19 ± 1.20	4.72 ± 0.92	...	...
990123	1.600	66.09 ± 6.33	128.31 ± 12.29	0.089	5.08 ± 0.49
990217	...	< 1.25	< 0.95	...	...
990510	1.619	41.07 ± 3.68	82.09 ± 7.35	0.054	1.20 ± 0.11
990627	...	4.41 ± 0.85	3.36 ± 0.65	...	...
990704	...	10.23 ± 3.34	7.80 ± 2.54	...	...
990705	0.840	3.11 ± 1.14	1.23 ± 0.45	0.096	0.06 ± 0.02
990806	...	8.46 ± 3.14	6.45 ± 2.39	...	...
990907	...	11.58 ± 6.95	8.82 ± 5.29	...	...
991014	...	4.54 ± 1.71	3.46 ± 1.30	...	...
991216	1.020	287.21 ± 14.73	183.22 ± 9.39	0.051	2.38 ± 0.12
000115	...	78.3 ± 14.12	59.67 ± 10.76	...	...
000210	0.846	4.56 ± 1.16	1.83 ± 0.47	...	...
000214	...	10.25 ± 2.16	7.81 ± 1.65	...	...
000528	...	2.61 ± 1.27	1.99 ± 0.97	...	...
000529	...	2.43 ± 0.47	1.85 ± 0.36	...	...
000926	2.037	20.41 ± 8.06	71.69 ± 28.31	0.140	7.01 ± 2.77
001025	...	67.85 ± 51.48	51.71 ± 39.22	...	...
001109	...	17.02 ± 2.06	12.97 ± 1.57	...	...
010214	...	3.95 ± 0.80	3.01 ± 0.61	...	...
010220	...	0.79 ± 0.21	0.61 ± 0.16	...	...
010222	1.477	86.50 ± 9.88	137.86 ± 15.75	0.080	4.41 ± 0.50
011211	2.14	2.23 ± 0.39	8.86 ± 1.56	...	...
020322	...	7.75 ± 0.67	5.91 ± 0.51	...	...
020405	0.698	68.98 ± 20.21	17.29 ± 5.07	0.285	6.98 ± 2.04
020813	1.254	113.98 ± 17.01	121.21 ± 18.09	0.066	2.61 ± 0.39
021004	2.323	13.50 ± 2.47	65.36 ± 11.95	0.240	18.7 ± 3.4

Note. — The columns are (left to right): (1) GRB name, (2) redshift, (3) X-ray flux at $t = 10$ hr, (4) X-ray luminosity at $t = 10$ hr, (5) jet opening angle, and (6) beaming-corrected X-ray luminosity at $t = 10$ hr.

CHAPTER 5

The Non-Relativistic Evolution of GRBs 980703 and 970508: Beaming-Independent Calorimetry

E. Berger[a], S. R. Kulkarni[a], & D. A. Frail[b]

[a]Department of Astronomy, 105-24 California Institute of Technology, Pasadena, CA 91125, USA

[b]National Radio Astronomy Observatory, P. O. Box 0, Socorro, NM 87801

Abstract

We use the Sedov-Taylor self-similar solution to model the radio emission from the γ-ray bursts (GRBs) 980703 and 970508, when the blastwave has decelerated to non-relativistic velocities. This approach allows us to infer the energy independent of jet collimation. We find that for GRB 980703 the kinetic energy at the time of the transition to non-relativistic evolution, $t_{\rm NR} \approx 40$ d, is $E_{\rm ST} \approx (1-6) \times 10^{51}$ erg. For GRB 970508 we find $E_{\rm ST} \approx 3 \times 10^{51}$ erg at $t_{\rm NR} \approx 100$ d, nearly an order of magnitude higher than the energy derived in Frail et al. (2000c). This is due primarily to revised cosmological parameters and partly to the maximum likelihood fit we use here. Taking into account radiative losses prior to $t_{\rm NR}$, the inferred energies agree well with those derived from the early, relativistic evolution of the afterglow. Thus, the analysis presented here provides a robust, geometry-independent confirmation that the energy scale of cosmological GRBs is about 5×10^{51} erg, and additionally shows that the central engine in these two bursts did not produce a significant amount of energy in mildly relativistic ejecta at late time. Furthermore, a comparison to the prompt energy release reveals a wide dispersion in the γ-ray efficiency, strengthening our growing understanding that E_γ is a not a reliable proxy for the total energy.

SECTION 5.1

Introduction

The two fundamental quantities in explosive phenomena are the kinetic energy, E_K, and the mass of the explosion ejecta, $M_{\rm ej}$, or equivalently the expansion velocity, $\beta \equiv v/c$, or Lorentz factor, $\Gamma = (1-\beta^2)^{-1/2}$. Together, these gross parameters determine the appearance and evolution of the resulting explosion. Gamma-ray bursts (GRBs) are distinguished by a highly relativistic initial velocity, $\Gamma_0 \gtrsim 100$, as inferred from their nonthermal prompt emission (Goodman 1986; Paczynski 1986). For the range of γ-ray isotropic-equivalent energies observed in GRBs, $E_{\gamma,\rm iso} \sim 10^{51} - 10^{54}$ erg (Bloom et al. 2001), this indicates $M_{\rm ej} \sim 10^{-5} - 10^{-3}$ M$_\odot$, compared to several M$_\odot$ in supernovae (SNe).

The true energy release of GRBs depends sensitively on the geometry of the explosion. For a collimated outflow ("jet") with a half-opening angle θ_j, it is $E = f_b E_{\text{iso}}$, where $f_b \equiv [1 - \cos(\theta_j)]$ is the beaming fraction; the true ejecta mass is also a factor of f_b lower. Over the past several years there has been growing evidence for such collimated outflows coming mainly from achromatic breaks in the afterglow light curves (e.g., Kulkarni et al. 1999a; Stanek et al. 1999). The epoch at which the break occurs, t_j, corresponds to the time at which the ejecta bulk Lorentz factor decreases below θ_j^{-1} (Rhoads 1999; Sari et al. 1999).

In this context, several studies have shown that the beaming-corrected energies of most GRBs, in both the prompt γ-rays and afterglow phase, are of the order of 10^{51} erg (Frail et al. 2001; Panaitescu & Kumar 2002; Berger et al. 2003a; Bloom et al. 2003b; Yost et al. 2003). The various analyses are sensitive to the energy contained in ejecta with different velocities, $\Gamma \gtrsim 100$ in the γ-rays, $\Gamma \gtrsim 10$ in the early X-rays, and $\Gamma \gtrsim$ few in the broad-band afterglow. However, *none* are capable of tracing the existence and energy of non-relativistic ejecta.

Frail et al. (2000c) overcame this problem in the case of GRB 970508 by modeling the afterglow radio emission in the non-relativistic phase, thus inferring $E_K \approx 5 \times 10^{50}$ erg. This analysis has two significant advantages. First and foremost it is independent of jet collimation since the blastwave approaches spherical symmetry on the same timescale that it becomes non-relativistic (Livio & Waxman 2000). Second, this analysis relies on the simple and well-understood Sedov-Taylor dynamics of spherical blastwaves, as opposed to the hydrodynamics of spreading relativistic jets. In addition, the peak of the synchrotron spectrum on the relevant timescale lies in the radio band where the afterglow is observable for several hundred days.

Two recent developments make similar analyses crucial. We now recognize that some GRBs are dominated by mildly relativistic ejecta (Berger et al. 2003c). For example, for GRB 030329 the kinetic energy inferred from the afterglow emission, $E_K(\Gamma \sim \text{few}) \approx 5 \times 10^{50}$ erg (Berger et al. 2003c), was an order of magnitude higher than the γ-ray energy release (Price et al. 2003). Similarly, for GRB 980425 $E_\gamma \approx 8 \times 10^{47}$ erg (Galama et al. 1998b; Pian et al. 2000) was about 1% of the relativistic kinetic energy of the associated SN 1998bw, $E_K \approx 10^{50}$ erg (Kulkarni et al. 1998; Li & Chevalier 1999). This begs the question, is there even more energy emerging from the engine, either at the time of the burst or later on, at non-relativistic velocities?

Second, there is a growing interest in "unification models" for GRBs, X-ray flashes (XRFs) and core-collapse SNe of type Ib/c, relying primarily on energetics arguments. For example, Lamb et al. (2004) argue that GRBs and XRFs share an energy scale of $\sim 10^{49}$ erg, and that all type Ib/c SNe give rise to GRBs or XRFs. Both conclusions result from significantly smaller values of θ_j compared to those inferred in the past, such that the energy scale, $\propto \theta_j^2$, is lower by a factor of ~ 100 and the true GRB rate, $\propto \theta_j^{-2}$, matches locally the type Ib/c SN rate. Given the important ramifications of the GRB energy scale for progenitor scenarios we would like to independently address the question: Is the energy scale of cosmic explosions 10^{49} erg, implicating all type Ib/c SNe in the production of GRBs, or does it cluster on $\sim 10^{51}$ erg?

The answer will also provide an independent confirmation of the jet paradigm by comparison to the isotropic-equivalent energies. This is crucial since other explanations for the light curve breaks have been suggested, including changes in the density of the circumburst medium, a transition to a non-relativistic evolution on the timescale of a few days (due to a high circumburst density), and changes in the energy spectrum of the radiating electrons (Dai & Lu 2001; Panaitescu 2001; Wei & Lu 2002).

Here we address the possibility of significant contribution from non-relativistic ejecta and robustly determine the energy scale of GRBs independent of geometrical assumptions, using Very Large Array[1] radio observations of the afterglows of GRBs 970508 and 980703 in the non-relativistic phase. We generally follow the treatment of Frail et al. (2000c), but unlike these authors we carry out a full least-squares fit to the data.

---

[1] The VLA is operated by the National Radio Astronomy Observatory, a facility of the National Science Foundation operated under cooperative agreement by Associated Universities, Inc.

SECTION 5.2

The Non-Relativistic Blastwave and Fireball Calorimetry

The dynamical evolution of an ultra-relativistic blastwave expanding in a uniform medium (hereafter, *ISM*) is described in terms of its Lorentz factor, $\Gamma = (17E_{\text{iso}}/8\pi n m_p c^2 r^3)^{1/2}$, where r is the radius of the blastwave and n is the number density of the circumburst medium (Blandford & McKee 1976). This, along with the relation for the observer time, which for the line of sight to the center of the blastwave is $t \approx r/8\Gamma^2 c$ (e.g., Sari 1997), determines the evolution of the radius and Lorentz factor. For a spherical blastwave the expansion will eventually become non-relativistic on a timescale[2], $t_{\text{NR}} \approx 65(E_{\text{iso},52}/n_0)^{1/3}$ d, determined by the condition that the mass swept up by the blastwave, $M_{\text{sw}} \approx E_{\text{iso}}/c^2$, where M_{sw}.

An initially collimated outflow becomes non-relativistic at $t_{\text{NR}} \approx 40(E_{\text{iso},52}/n_0)^{1/4}t_{j,d}^{1/4}$ d (Livio & Waxman 2000). Moreover, as the jet expands sideways (at $t \gtrsim t_j$) the outflow approaches spherical symmetry on a timescale, $t_s \approx 150(E_{\text{iso},52}/n_0)^{1/4}t_{j,d}^{1/4}$ d, similar to t_{NR}. Thus, regardless of the initial geometry of the outflow the non-relativistic expansion is well-approximated as a spherical outflow. We note that this discussion can be generalized to a range of radial density profiles. Here, in addition to the *ISM* model, we focus on a density profile, $\rho = Ar^{-2}$ (hereafter, *Wind*), appropriate for mass loss with a constant rate, $\dot{M}_w$, and speed, v_w (Chevalier & Li 2000).

Following the transition to non-relativistic expansion, the dynamical evolution of the blastwave is described by the Sedov-Taylor self-similar solution (Sedov 1946; von Neumann 1947; Taylor 1950). In this case the radius of the shock is given by $r \propto (E_{\text{ST}}t^2/A)^{1/(5-s)}$, with $\rho = Ar^{-s}$. Thus, in the *ISM* case $r \propto (E_{\text{ST}}t^2/nm_p)^{1/5}$, while in the *Wind* case $r \propto (E_{\text{ST}}t^2/A)^{1/3}$. The constant of proportionality, $\xi(\hat{\gamma})$, depends on the adiabatic index of the gas, $\hat{\gamma}$, and is equal to 1.05 in the *ISM* case and 0.65 in the *Wind* case for $\hat{\gamma} = 13/9$. The latter is appropriate for pressure equilibrium between relativistic electrons and non-relativistic protons[3] (Frail et al. 2000c). The circumburst material shocked by the blastwave is confined downstream to a thin shell of width r/η, with $\eta \approx 10$.

To calculate the synchrotron emission emerging from this shock-heated material we make the usual assumptions. First, the relativistic electrons are assumed to obey a power-law distribution, $N(\gamma) \propto \gamma^{-p}$ for $\gamma \geq \gamma_m$. Second, the energy densities in the magnetic field and electrons are assumed to be a non-varying fraction (ϵ_B and ϵ_e, respectively) of the shock energy density. Coupled with the synchrotron emissivity and taking into account self-absorption, the flux received by an observer at frequency ν and time t is given by (e.g., Frail et al. 2000c):

$$F_\nu = F_0(t/t_0)^{\alpha_F}[(1+z)\nu]^{5/2}(1-e^{-\tau})f_3(\nu/\nu_m)f_2^{-1}(\nu/\nu_m), \tag{5.1}$$

the optical depth is given by

$$\tau_\nu = \tau_0(t/t_0)^{\alpha_\tau}[(1+z)\nu]^{-(p+4)/2}f_2(\nu/\nu_m), \tag{5.2}$$

and the function

$$f_l(x) = \int_0^x F(y)y^{(p-l)/2}dy. \tag{5.3}$$

Here, $\nu_m = \nu_0(t/t_0)^{\alpha_m}/(1+z)$ is the synchrotron peak frequency corresponding to electrons with $\gamma = \gamma_m$, $F(y)$ is given in, e.g., Rybicki & Lightman (1979), and the temporal indices α_F, α_τ and α_m are determined by the density profile of the circumburst medium. In the *ISM* case $\alpha_F = 11/10$, $\alpha_\tau = 1 - 3p/2$, and $\alpha_m = -3$, while in the *Wind* case $\alpha_F = 11/6$, $\alpha_\tau = -1 - 7p/6$, and $\alpha_m = -7/3$ (Waxman 2004b). Equations 5.1–5.3 include the appropriate redshift transformations to the rest-frame of the burst.

---

[2] Here and throughout the paper we use the notation $q = 10^x q_x$.

[3] The relative pressure between the protons and relativistic electrons depends on the fraction of energy in relativistic electrons, ϵ_e. If this fraction is low, the pressure may be dominated by the non-relativistic protons in which case $\hat{\gamma} = 5/3$. As we show below, ϵ_e for both GRB 980703 and GRB 970508 is in the range ~ 0.1 to 0.5 and thus $\hat{\gamma} = 13/9$ is applicable.

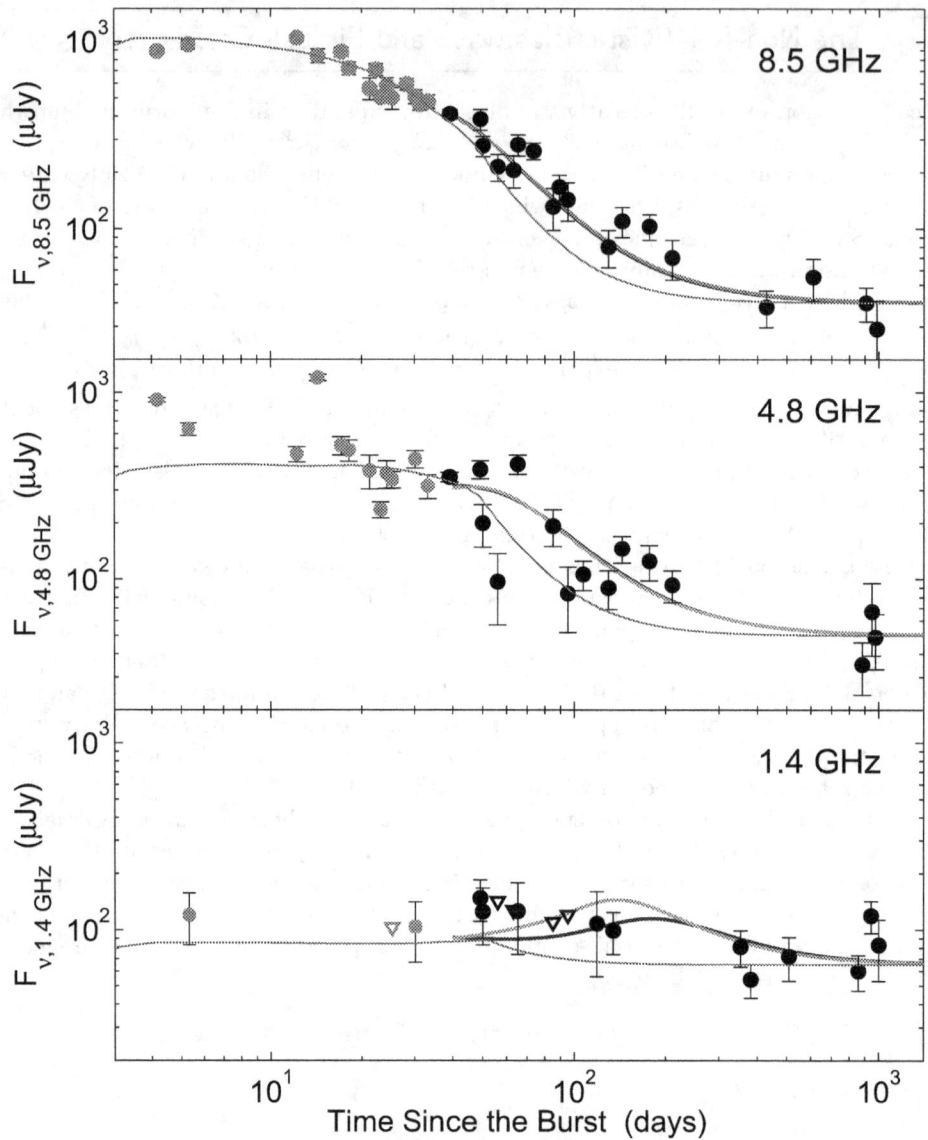

Figure 5.1: Radio light curves of the afterglow of GRB 980703 at 1.4, 4.9 and 8.5 GHz. Only data at $t \geq t_{\mathrm{NR}} = 40$ d (black circles) are used in the fit. The data exhibit a clear flattening relative to the relativistic evolution of the afterglow (thin gray line; Frail et al. 2003b) in agreement with the expected change from $F_\nu \propto t^{-p}$ (jet) to $F_\nu \propto t^{(21-15p)/10}$ (*ISM*) or $F_\nu \propto t^{(5-7p)/6}$ (*Wind*) in the non-relativistic regime. The best-fit light curves for the *ISM* (black) and *Wind* (gray) models are indistinguishable. The models include a contribution from the host galaxy of 40, 50 and 65 μJy at 8.5, 4.9 and 1.4 GHz, respectively.

Based on the temporal scalings the synchrotron flux in the optically-thin regime ($\nu \gg \nu_m, \nu_a$) evolves as $F_\nu \propto t^{(21-15p)/10}$ (*ISM*) or $F_\nu \propto t^{(5-7p)/6}$ (*Wind*); here the synchrotron self-absorption frequency, ν_a, is defined by the condition $\tau_\nu(\nu_a) = 1$. Thus, for $\nu \gg \nu_m, \nu_a$ the transition to non-relativistic expansion is manifested as a steepening of the light curves at t_{NR} if the outflow is spherical (Sari et al. 1998; Chevalier & Li 2000), or a flattening if the outflow was initially collimated (Sari et al. 1999). Below, we use this behavior to estimate $t_{\rm NR}$ for GRBs 980703 and 970508/

In §5.3 and §5.4 we use the temporal decay indices and Equations 5.1–5.3 to carry out a least-squares fit to the data at $t > t_{\rm NR}$ with the free parameters F_0, τ_0, ν_0 and p. These parameters are in turn used to calculate the physical parameters of interest, namely r, n_e, γ_m and B; $n_e \approx (\eta/3)n$ is the shocked electron density (Frail et al. 2000c). Since only three spectral parameters are available, this leaves the radius unconstrained and thus,

$$B = 11.7(p+2)^{-2}F_{0,-52}^{-2}(r_{17}/d_{28})^4 \ {\rm G}, \tag{5.4}$$

$$\gamma_m = 6.7(p+2)F_{0,-52}\nu_{0,9}^{1/2}(r_{17}/d_{28})^{-2}, \tag{5.5}$$

$$n_e = 3.6 \times 10^{10} c_n \eta_1 F_{0,-52}^3 \nu_{0,9}^{(1-p)/2}\tau_{0,32}^{-1}r_{17}^{-1}(r_{17}/d_{28})^{-6} \ {\rm cm}^{-3}, \tag{5.6}$$

$$c_n = (1.67 \times 10^3)^{-p}(5.4 \times 10^2)^{(1-p)/2}(p+2)^2/(p-1). \tag{5.7}$$

In the *Wind* model, the density is appropriate at $r_{\rm ST} \equiv r(t_{\rm NR})$, i.e., $\rho(r) = nm_p(r/r_{\rm NR})^{-2}$.

To determine the radius of the blastwave a further constraint is needed. We note that the energy contained in the electrons and magnetic field cannot exceed the thermal energy of the Sedov-Taylor blastwave, which accounts for about half of the total energy (Frail et al. 2000c). The energy in the electrons is given by $E_e = [(p-1)/(p-2)]n_e\gamma_m m_e c^2 V$, while the energy in the magnetic field is $E_B = B^2 V/8\pi$; here $V = 4\pi r^3/\eta$ is the volume of the synchrotron emitting shell. Thus, using Equations 5.4–5.7 and the condition $E_e + E_B \leq E_{\rm ST}/2$ we can constrain the range of allowed values of r. In the *ISM* model $E_{\rm ST} = nm_p(r/1.05)^5[t_{\rm NR}/(1+z)]^{-2}$, while in the *Wind* model $E_{\rm ST} = A(r/0.65)^3[t_{\rm NR}/(1+z)]^{-2}$.

With a constraint on the radius we can also ensure self-consistency by calculating the velocity of the blastwave when it enters the Sedov-Taylor phase, $v_{\rm ST} = 2r(1+z)/5t_{\rm NR}$ (*ISM*) or $v_{\rm ST} = 2r(1+z)/3t_{\rm NR}$ (*Wind*). We expect that roughly $v \sim c$. Finally, the isotropic-equivalent mass of the ejecta is given by $M_{\rm ej} = 4\pi nm_p r_{\rm ST}^3$ (*ISM*) or $4\pi A r_{\rm ST}$ (*Wind*). The actual ejecta mass is reduced by a factor f_b relative to this value.

SECTION 5.3

GRB 980703

In Figure 5.1 we plot the radio light curves of GRB 980703. The data are taken from Berger et al. (2001b) and Frail et al. (2003b). Two gross changes in the light curves evolution are evident: a flattening at $t \approx 40$ d at 4.9 and 8.5 GHz and a transition to a constant flux density at late time. The latter is due to radio emission from the host galaxy of GRB 980703 with flux densities at 1.4, 4.9 and 8.5 GHz of 65, 50 and 40 μJy, respectively (Berger et al. 2001b). The flattening at $t \approx 40$ d marks the transition to non-relativistic evolution following a period of sideways expansion of the initially collimated outflow (Figure 5.1). A similar value, $t_{\rm NR} \approx 30 - 50$ d has been inferred by Frail et al. (2003b) from tracking the evolution of the blastwave Lorentz factor in the relativistic phase. We therefore use here $t_{\rm NR} = 40$ d.

We follow the method outlined in §5.2 using both the *ISM* and *Wind* cases. The results of both fits, shown in Figure 5.1, are overall indistinguishable. In what follows we quote the results of the *ISM* model. The best-fit parameters ($\chi_{\rm min}^2 = 123$ for 45 degrees of freedom) are: $F_{0,-52} \approx 2.7$, $\tau_{0,32} \approx 80$, $\nu_{0,9} \approx 4.6$ and $p \approx 2.8$. The relatively large values of $\chi_{\rm min}^2$ is due primarily to fluctuations induced by interstellar scintillation, particularly at 4.9 GHz.

Using $d_{28} = d_{L,28}/(1+z)^{1/2} = 1.4$ ($z = 0.966$, $H_0 = 71$ km s^{-1} Mpc^{-1}, $\Omega_m = 0.27$ and $\Omega_\Lambda = 0.73$), and Equations 5.4–5.7 we find $B \approx 1.8 \times 10^{-2}r_{17}$ G, $\gamma_m \approx 300r_{17}$, and $n_e \approx 4.9 \times 10^3 r_{17}^{-7}$ cm^{-3}. From these parameters we calculate $E_e \approx 3.4 \times 10^{51}r_{17}^{-6}$ erg, $E_B \approx 1.7 \times 10^{46}r_{17}^{11}$ erg, and $E_{\rm ST} \approx 6.2 \times 10^{51}r_{17}^{-2}$. These results are plotted in Figures 5.2 and 5.3.

The range of blastwave radii allowed by the constraint $E_e + E_B \lesssim E_{\rm ST}/2$ is $r_{17} \approx 1.05 - 2.5$, resulting in a range of values for the Sedov-Taylor energy, $E_{\rm ST} \approx (1 - 6) \times 10^{51}$ erg. Given the strong

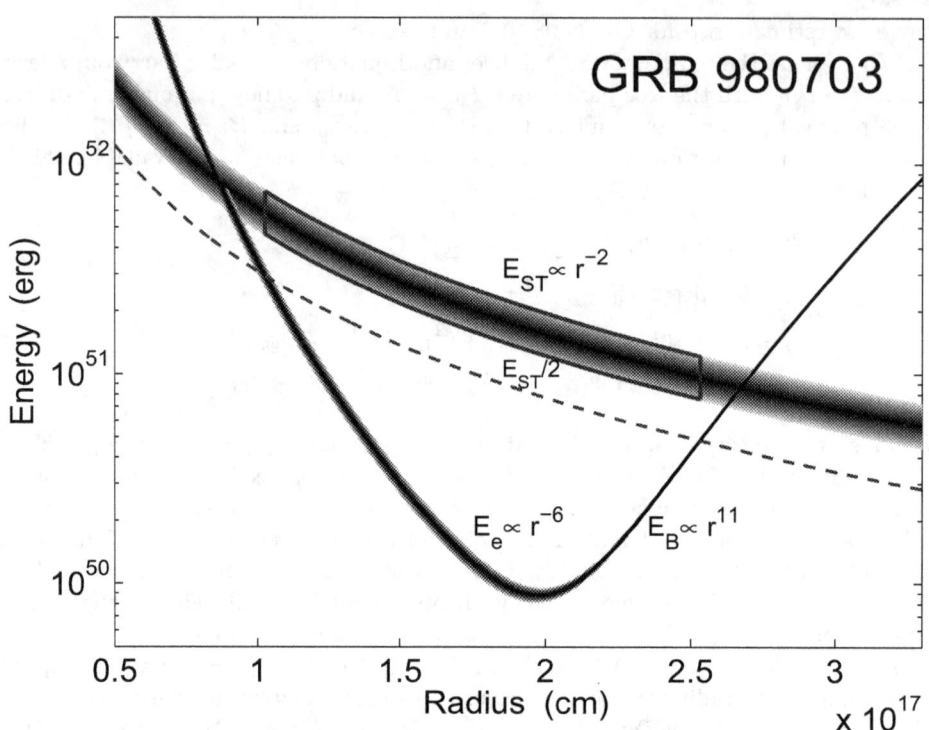

Figure 5.2: Energies associated with the afterglow of GRB 980703 in the non-relativistic Sedov-Taylor phase as a function of the (unconstrained) blastwave radius. The thin curve is the sum of the energy in relativistic electron ($E_e \propto r^{-6}$) and in the magnetic fields ($E_B \propto r^{11}$). Also plotted are the Sedov-Taylor energy ($E_{ST} \propto r^{-2}$) and the thermal component, $E_{ST}/2$. The shading corresponds to an uncertainty of 30% in the value of the synchrotron frequency ν_0 at $t = t_{NR}$. The value of $E_{ST}/2$ provides an additional constraint, $E_e + E_B \leq E_{ST}/2$, which limits the range of allowed radii in the solution (boxed region).

dependence on radius, the ratio of energy in the electrons to the energy in the magnetic field ranges from $\epsilon_e/\epsilon_B \approx 0.03 - 9 \times 10^4$, while the specific values range from $\epsilon_e \approx 0.01 - 0.45$ and $\epsilon_B \approx 5 \times 10^{-6} - 0.4$. The circumburst density is in the range $n \approx 8 - 3.5 \times 10^3$ cm^{-3}, while the blastwave velocity is $\beta_{ST} \approx 0.8 - 1.9$. Finally, the isotropic-equivalent mass of the ejecta ranges from $(1 - 40) \times 10^{-4}$ M$_\odot$.

A comparison to the values derived by Frail et al. (2003b) using modeling of the afterglow emission in the relativistic phase is useful. These authors find $n \approx 30$ cm^{-3}, $\epsilon_e \approx 0.27$ and $\epsilon_B \approx 2 \times 10^{-3}$. Using the same density in our model (Figure 5.3), as required by the *ISM* density profile, gives a radius $r_{17} \approx 1.75$ and hence $\epsilon_e \approx 0.06$ and $\epsilon_B \approx 4 \times 10^{-3}$, in rough agreement; the energy is $E_{ST} \approx 2 \times 10^{51}$ erg.

If we assume alternatively that the energy in relativistic electrons and the magnetic field are in equipartition, we find $r_{17} \approx 2.05$. In this case, $E_{ST} \approx 1.5 \times 10^{51}$ erg, $n \approx 10$ cm^{-3}, $B \approx 0.3$ G, and $\epsilon_e = \epsilon_B = 0.03$.

SECTION 5.4

GRB 970508

The non-relativistic evolution of GRB 970508 was studied by Frail et al. (2000c). These authors provide a rough model for the radio emission beyond $t_{NR} \approx 100$ d and argue that the constraint $E_e + E_B \lesssim E_{ST}/2$ requires the electron and magnetic field energy to be in equipartition, $\epsilon_e = \epsilon_B \approx 0.25$, with

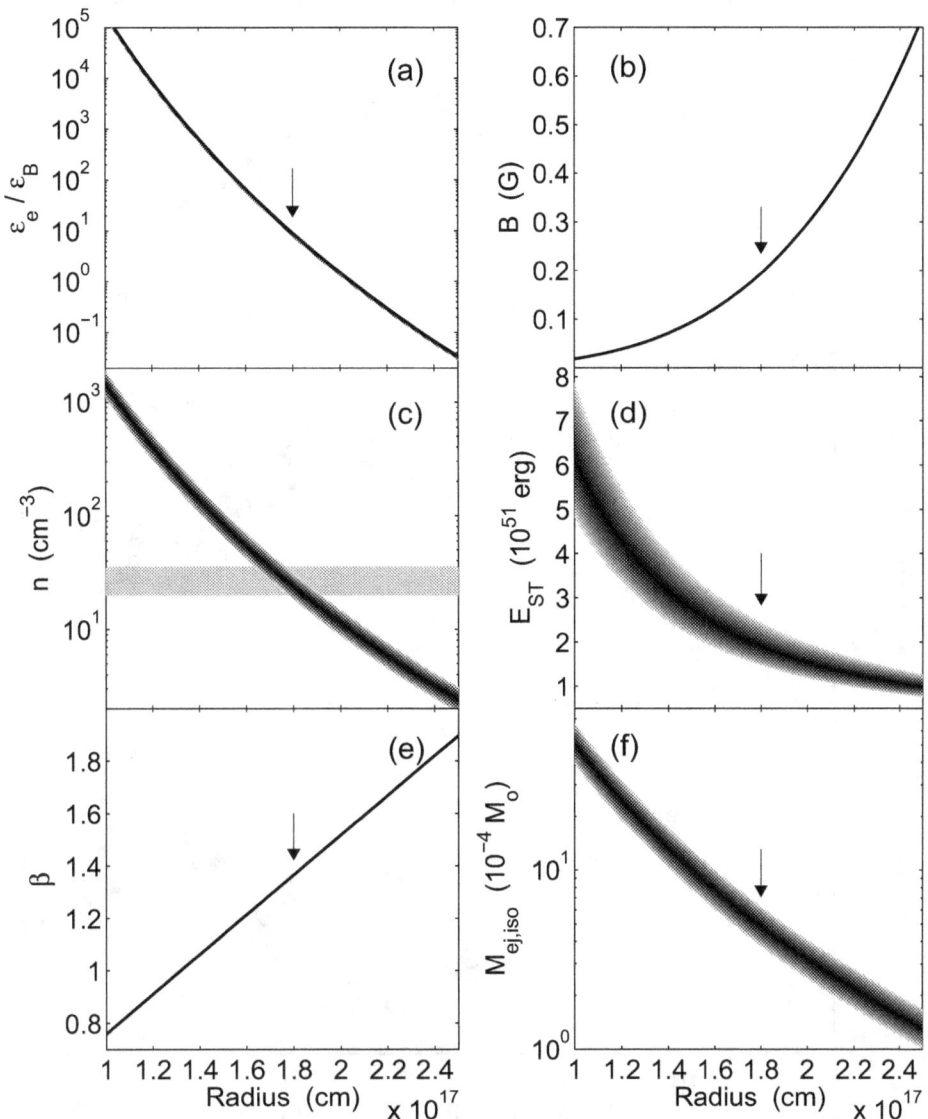

Figure 5.3: Physical parameters of the Sedov-Taylor blastwave for GRB 980703 at $t_{\mathrm{NR}} = 40$ d for the range of radii that obey the constraint $E_e + E_B \leq E_{\mathrm{ST}}/2$ (Figure 5.2): (a) The ratio of energy in the relativistic electrons to that in the magnetic fields, (b) the magnetic field strength, (c) the density of the circumburst medium, (d) the Sedov-Taylor energy, (e) the velocity of the blastwave, and (f) the isotropic-equivalent mass of the ejecta produced by the central engine and responsible for the afterglow emission. The light shaded region in (c) marks the range of densities inferred from the relativistic evolution of the fireball, $n \approx 20 - 35$ cm^{-3} (Frail et al. 2003b). With the additional constraint that the density derived here conform to this value, we derive the values of ϵ_e/ϵ_B, B, E_{ST}, β, and M_{ej} marked by arrows.

$E_{\mathrm{ST}} \approx 4.4 \times 10^{50}$ erg. Here we perform a full least-squares fit, using $t_{\mathrm{NR}} = 100$ d, and find somewhat different results. We use $t_{\mathrm{NR}} \approx 100$ d, noting that for GRB 970508 the outflow appears to be weakly-collimated (Yost et al. 2003), and hence the transition is manifested as a mild steepening of the light

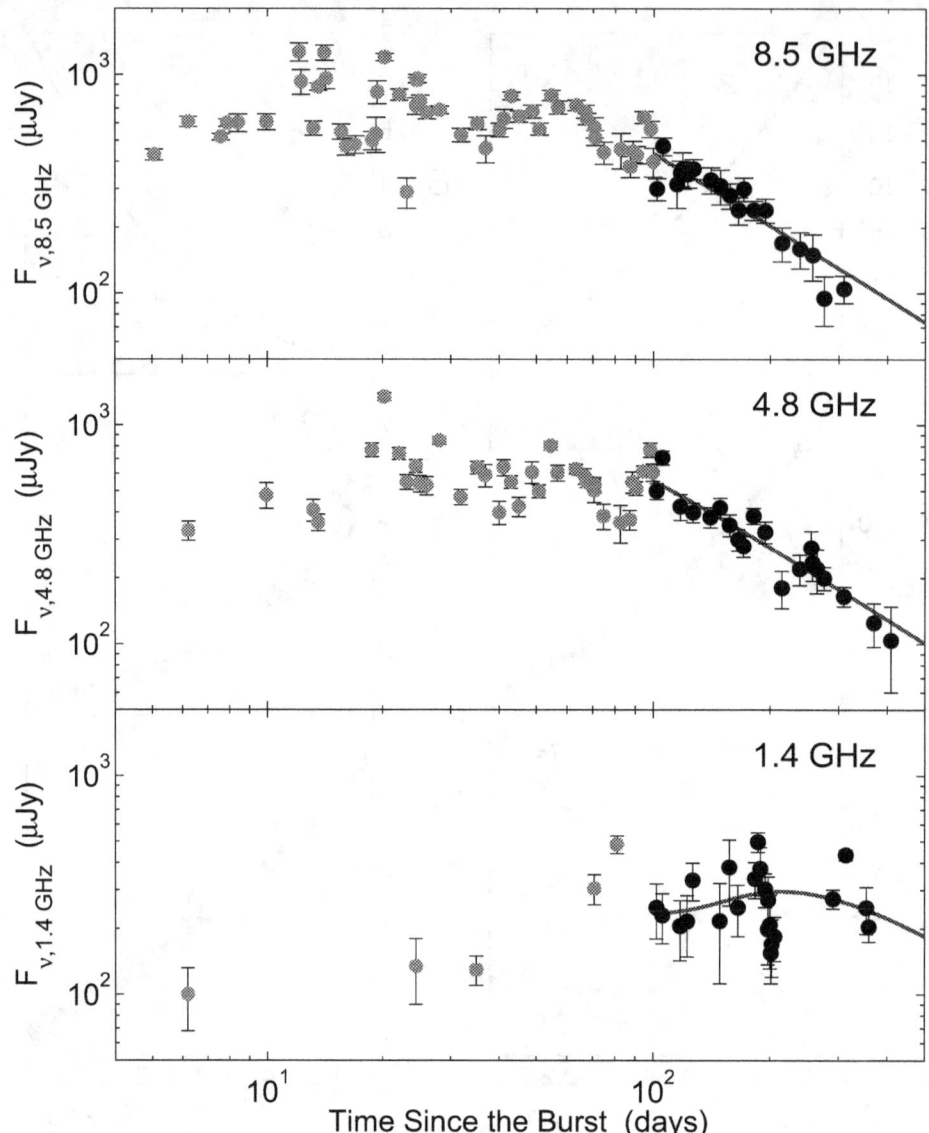

Figure 5.4: Radio light curves of the afterglow of GRB 970508 at 1.4, 4.9 and 8.5 GHz. Only data at $t \geq t_{\mathrm{NR}} = 100$ d (black circles) are used in the fit. The best-fit light curves for the *ISM* model are shown (black); the *Wind* model can be ruled out since it requires $p < 2$.

curves (see §5.2).

The best-fit parameters in the *ISM* model[4] ($\chi^2_{\mathrm{min}} = 164$ for 58 degrees of freedom) are: $F_{0,-52} \approx 38$, $\tau_{0,32} \approx 3.1 \times 10^{-3}$, $\nu_{0,9} \approx 3$ and $p \approx 2.17$. The large value of χ^2_{min} is due primarily to interstellar scintillation.

In comparison, Frail et al. (2000c) use $F_{0,-52} \approx 41$, $\tau_{0,32} \approx 5.3 \times 10^{-3}$, $\nu_{0,9} \approx 9.5$, and they set $p = 2.2$; a solution with $\nu_{0,9} \approx 4.2$ is also advocated but it is not used to derive the physical parameters

---

[4] We do not consider the *Wind* case since in this model the observed decay rates at 4.9 and 8.5 GHz, $F_\nu \propto t^{-1.2}$, require $p \approx 1.7$ and hence an infinite energy. This can be avoided by assuming a break in the electron energy distribution at $\gamma_b > \gamma_m$ with a power law index $q > 2$, but we do not have the data required to constrain either γ_b or q.

of the blastwave. The formal χ^2 values for these solutions are 225 and 254, respectively, somewhat worse than the solution found here.

As a result, we find that solutions away from equipartition are allowed. Adopting the cosmological parameters used by Frail et al. (2000c), $H_0 = 70$ km s^{-1} Mpc^{-1}, $\Omega_m = 1$ and $\Omega_\Lambda = 0$, we find $E_{ST} \approx (6-11) \times 10^{50}$ erg, a factor of about $20-100\%$ higher than the values inferred by these authors.

Using the currently favored cosmology (§5.3), we find instead that the distance to the burst is higher by about 30%, $d_{28} = 1.21$ compared to 0.94 (Frail et al. 2000c). The change in distance has a significant effect on the derived parameters since $E_e \propto d^8$, $E_B \propto d^{-8}$ and $E_{ST} \propto n \propto d^6$. Thus, we find that the constraint on $E_e + E_B$ indicates $r_{17} \approx 3.7 - 5.9$ and therefore, $B \approx 0.04 - 0.25$ G, $\gamma \approx 65 - 165$ and $n \approx 0.4 - 10$ cm^{-3}. The Sedov-Taylor energy is $E_{ST} \approx (1.5 - 3.8) \times 10^{51}$ erg, while $\epsilon_e \approx 0.07 - 0.5$ and $\epsilon_B \approx 0.001 - 0.45$ (Figures 5.5 and 5.6). Assuming equipartition, we find $r_{17} = 5.3$, $E_{ST} = 1.8 \times 10^{51}$ erg, and $\epsilon_e = \epsilon_b = 0.11$. The derived energy is about a factor of four higher than the previous estimate (Frail et al. 2000c).

A comparison of our best-fit model with the flux of the afterglow in the optical R-band at $t = 110$ d, $F_{\nu,R} \approx 0.3$ μJy (Garcia et al. 1998), indicates a break in the spectrum. If we interpret this break as due to the synchrotron cooling frequency, above which the spectrum is given by $F_\nu \propto \nu^{-p/2}$, we find $\nu_c \approx 6 \times 10^{13}$ Hz. Since $\nu_c = 1.9 \times 10^{10} B^{-3} (t/110\,\mathrm{d})^{-2}$ Hz we infer $B \approx 0.073$ G and hence $r_{17} = 4.3$, $E_{ST} = 2.8 \times 10^{51}$, $\epsilon_e = 0.25$ and $\epsilon_B = 8 \times 10^{-3}$. These values are in rough agreement with those inferred from modeling of the relativistic phase (Panaitescu & Kumar 2002; Yost et al. 2003), although our value of ϵ_B is somewhat lower.

SECTION 5.5

Radiative Corrections

The energies derived in §5.3 and §5.4 are in fact lower limits on the initial kinetic energy of the blastwave due to synchrotron radiative losses. These play a role primarily in the fast-cooling regime ($\nu_c \ll \nu_m$), which dominates in the early stages of the afterglow evolution (e.g., Sari et al. 1998).

Yost et al. (2003) estimate the time at which fast-cooling ends, $t_{cm} \approx 0.1$ and 1.4 days after the burst for GRB 970508 and GRB 980703, respectively. Using these values, and our best estimate of $\epsilon_e \approx 0.06$ (980703) and $\epsilon_e \approx 0.25$ (970508), we calculate the radiative corrections, $E \propto t^m$, going back from t_{NR} to about 90 s after the burst. Here $m \approx -17\epsilon/12$, with $\epsilon = \epsilon_e/(1 + 1.05\epsilon_e)$ for $t < t_{cm}$ and it is quenched by a factor $(\nu_m/\nu_c)^{(p-2)/2} < 1$ at later times. Thus, at low values of ϵ_e the radiative losses are negligible. The cutoff at 90 s corresponds to the approximate deceleration time of the ejecta, $t_{dec} \approx 90(E_{52}/n_0\Gamma_2^8)^{1/3}$ s.

We find that approximately 50% and 90% of the energy was radiated away before t_{NR} for GRBs 980703 and 970508, respectively. Thus, the initial kinetic energies are estimated to be 4×10^{51} erg and 3×10^{52} erg, respectively. The corrections from t_{NR} back to t_{cm}, 10% for GRB 980703 and 70% for GRB 970508, indicate $E_K \approx 2 \times 10^{51}$ and 9×10^{51} erg, respectively. Both estimates of the energy are in excellent agreement with those inferred from the relativistic evolution of the fireball at t_{cm} (Yost et al. 2003), $E_K \approx 3 \times 10^{51}$ erg (980703) and $E_K \approx 1.2 \times 10^{52}$ erg (970508).

SECTION 5.6

Discussion and Conclusions

Analysis of the synchrotron emission from a GRB blastwave in the non-relativistic phase has the advantage that it is independent of geometry and is described by the well-understood Sedov-Taylor self-similar solution. Using this approach to model the late-time radio emission from GRBs 980703 ($t > 40$ d) and 970508 ($t > 100$ d) we infer kinetic energies in the range $(1 - 6) \times 10^{51}$ erg and $(1.5 - 4) \times 10^{51}$ erg, respectively. Including the effect of radiative losses starting at $t_{dec} \sim 90$ s, we find that the initial kinetic energies were about 4×10^{51} erg and 3×10^{52} erg, respectively.

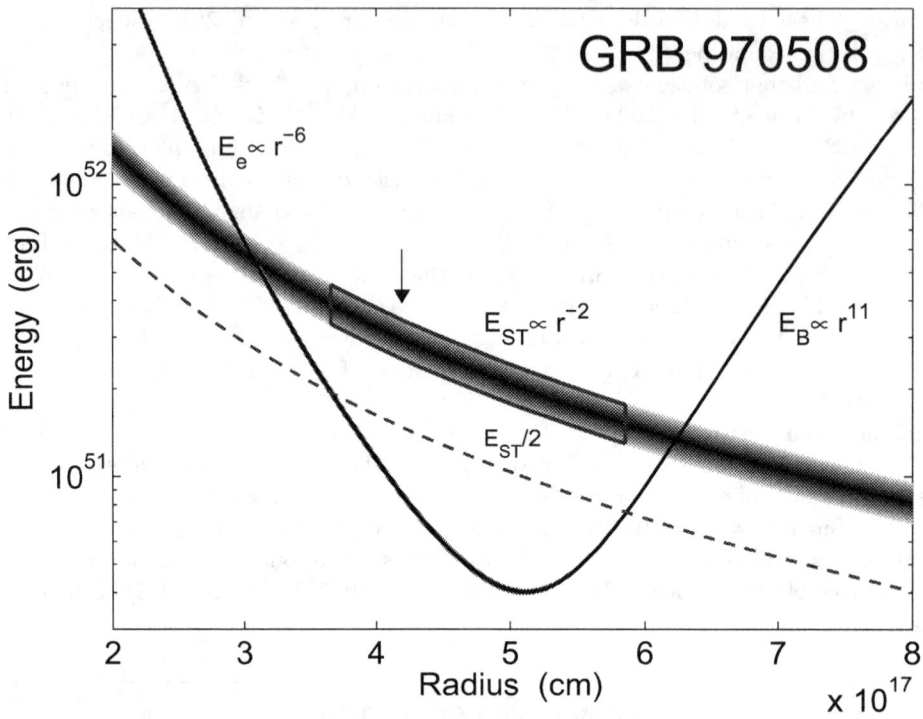

Figure 5.5: Energies associated with the afterglow of GRB 970508 in the non-relativistic Sedov-Taylor phase as a function of the (unconstrained) blastwave radius. The thin curve is the sum of the energy in relativistic electron ($E_e \propto r^{-6}$) and in the magnetic fields ($E_B \propto r^{11}$). Also plotted are the Sedov-Taylor energy ($E_{ST} \propto r^{-2}$) and the thermal component, $E_{ST}/2$. The shading corresponds to an uncertainty of 30% in the value of the synchrotron frequency ν_0 at $t = t_{NR}$. The value of $E_{ST}/2$ provides an additional constraint, $E_e + E_B \leq E_{ST}/2$, which limits the range of allowed radii in the solution (boxed region). Finally, the arrow marks the most likely solution using the value of the cooling frequency as estimated from a combination of the radio and optical data (§5.4). This additional parameter breaks the radius degeneracy, indicating $r \approx 4.2 \times 10^{17}$ cm and $E_{ST} \approx 3 \times 10^{51}$ erg

The inferred kinetic energies confirm, independent of any assumptions about the existence or opening angles of jets, that the energy scale of GRBs is $\sim 5 \times 10^{51}$ erg. We therefore unambiguously rule out the recent claim of Lamb et al. (2004) that the energy scale of GRBs is of the order of 10^{49} erg. Since the claimed low energies were based on the apparent correlation between $E_{\gamma,\mathrm{iso}}$ and the energy at which the prompt emission spectrum peaks, E_{peak} (Amati et al. 2002), we conclude that this relation, and the prompt emission in general, does not provide a reliable measure of the total energy. As a corollary, we rule out the narrow jet opening angles used by Lamb et al. (2004), $\theta_j \sim 0.1°$ and thus confirm that the true GRB rate is significantly lower than the rate of type Ib/c SNe (Berger et al. 2003b).

Finally, the overall agreement between the energies derived here and those inferred from modeling of the relativistic phase of the afterglow indicates that the central engine in GRBs 980703 and 970508 did not produce a significant amount of energy in mildly relativistic ejecta ($\Gamma\beta \gtrsim 2$) at late time, $t \sim t_{\mathrm{NR}}$. However, a comparison to the beaming-corrected γ-ray energies (Bloom et al. 2003b), $E_\gamma \approx 1.1 \times 10^{51}$ erg (GRB 980703) and $E_\gamma \sim 10^{51}$ erg (GRB 970508) reveals that the efficiency of the blastwave in producing γ-rays, ϵ_γ, varies considerably: $\sim 20\%$ for GRB 980703, but only $\sim 3\%$ for GRB 970508. The wide dispersion in ϵ_γ strengthens the conclusion that E_γ is not a reliable tracer of the total energy (Berger et al. 2003c).

The low value of ϵ_γ for GRB 970508 may indicate an injection of energy from mildly relativistic

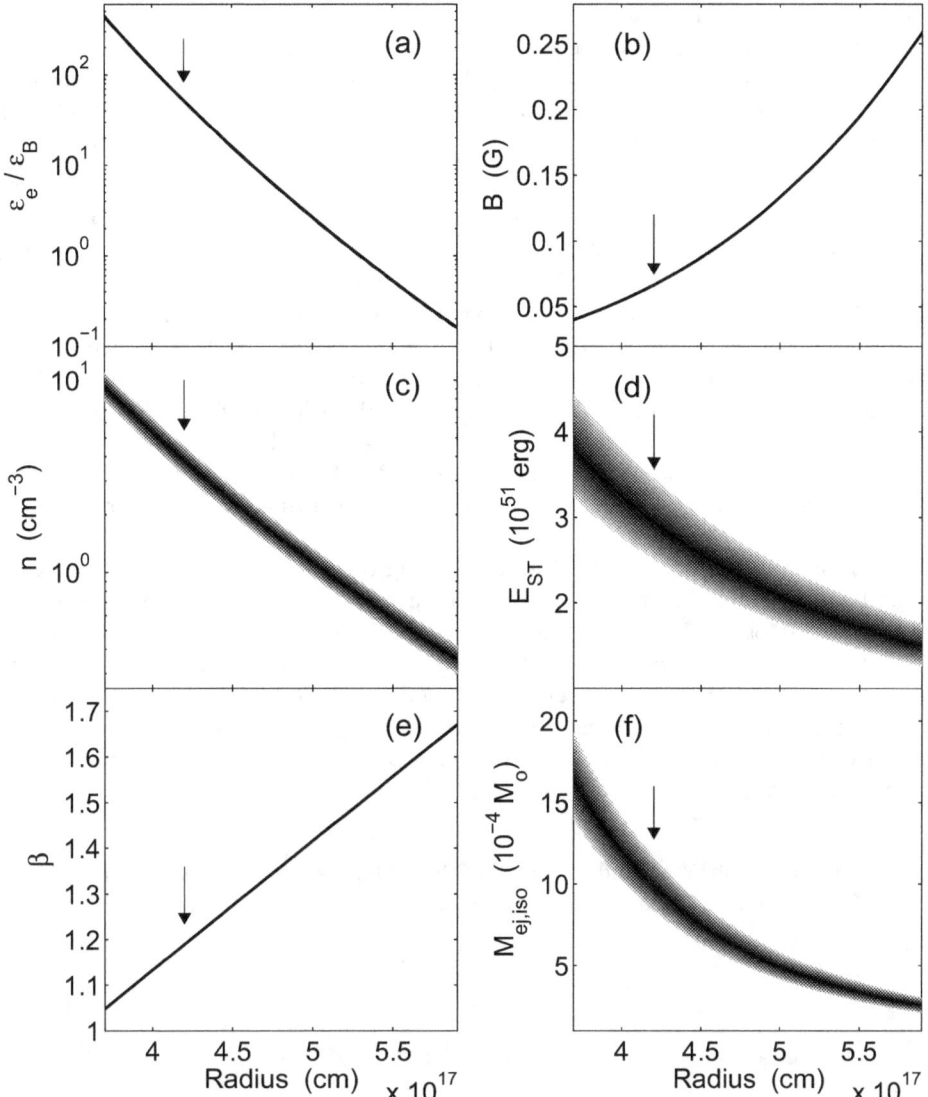

Figure 5.6: Physical parameters of the Sedov-Taylor blastwave for GRB 970508 at $t_{\mathrm{NR}} = 100$ d for the range of radii that obey the constraint $E_e + E_B \leq E_{\mathrm{ST}}/2$ (Figure 5.5): (a) The ratio of energy in the relativistic electrons to that in the magnetic fields, (b) the magnetic field strength, (c) the density of the circumburst medium, (d) the Sedov-Taylor energy, (e) the velocity of the blastwave, and (f) the isotropic-equivalent mass of the ejecta produced by the central engine and responsible for the afterglow emission. The arrows mark the most likely values using an estimate of the cooling frequency from a combination of the radio and optical data (§5.4 and Figure 5.5).

ejecta at early time. Both the optical and X-ray light curves of this burst exhibited a sharp increase in flux approximately 1 day after the burst, by a factor of about 4 and $\gtrsim 2$, respectively (Piro et al. 1998; Sokolov et al. 1998). The flux in these bands depends on energy as $F_\nu \propto E^{(p+3)/4}$ and $\propto E^{(p+2)/4}$, respectively (Sari et al. 1998). Thus, if we interpret the flux increase as due to injection of energy from ejecta with $\Gamma \sim 5 - 10$ (Panaitescu et al. 1998) we find an energy increase of about a factor of three. The analysis performed here provides an estimate of the total energy following the injection and thus

ϵ_γ appears to be low. The actual value of ϵ_γ is thus $\sim 10\%$.

Although GRBs 980703 and 970508 are currently the only bursts with sufficient radio data to warrant the full Sedov-Taylor analysis, flattening of radio light curves at late time have been noted in several other cases, most notably GRBs 980329, 991208, 000301C, 000418 and 000926 (Frail et al. 2004). Interpreting the flattening as a transition to non-relativistic expansion and using the expression for the flux at 8.5 GHz at the time of the transition, $F_\nu(t_{\mathrm{NR}}) \approx 50[(1+z)/2]^{-1/2}\epsilon_{e,-1}\epsilon_{B,-1}^{3/4}n_0^{3/4}E_{51}d_{28}^{-2}$ μJy (Livio & Waxman 2000), we find the rough results $n_0^{3/4}E_{51} \approx 6$ (980329), ≈ 4 (991208), ≈ 25 (000301C), ≈ 6 (000418), and ≈ 22 (000926). Thus, for typical densities, $\sim 1 - 10$ cm^{-3} (Panaitescu & Kumar 2002; Yost et al. 2003), the inferred kinetic energies are again of the order of $10^{51} - 10^{52}$ erg.

This leads to the following conclusions. First, the energy scale of cosmological bursts is about 5×10^{51} erg, at least three orders of magnitude higher than the kinetic energies in fast ejecta determined for local type Ib/c SNe from radio observations (Berger et al. 2002b, 2003b), and an order of magnitude higher relative to the nearby ($d \approx 40$ Mpc) GRB 980425 associated with SN 1998bw (Kulkarni et al. 1998; Li & Chevalier 1999; Waxman 2004a) and GRB 031203 ($z = 0.105$; Prochaska et al. 2004; Soderberg et al. 2004). Second, as already noted in the case of GRB 030329 (Berger et al. 2003c), there is a wide dispersion in the fraction of energy in ultra-relativistic ejecta, such that the γ-rays are a poor proxy for the total energy produced by the engine.

Thus, radio calorimetry is uniquely suited for addressing the relation between various cosmic explosions. So far, such studies reveal a common energy scale in relativistic ejecta of about 5 foe (foe $\equiv 10^{51}$ erg) for cosmological GRBs (Berger et al. 2003c), about 0.1 foe for the low redshift bursts (980425, 031203), and $\lesssim 10^{-3}$ foe in fast ejecta for type Ib/c SNe. The open question now is whether we are beginning to trace a continuum in the energetics of cosmic explosions, or whether the various classes truly represent distinct physical mechanisms with different energy scales. Fortunately, the best example to date of an object possibly bridging the various populations, GRB 030329, still shines brightly in the radio a year after the burst.

We thank Eli Waxman, Sarah Yost and Re'em Sari for valuable discussions, and the referee, Roger Chevalier, for useful comments. We acknowledge NSF and NASA grants for support.

Table 5.1.   Physical Parameters of GRBs 980703 and 970508

Parameter	GRB 980703	GRB 970508
r (10^{17} cm)	$1.05 - 2.5$	$3.7 - 5.9$
B (G)	$0.02 - 0.7$	$0.04 - 0.25$
γ	$8 - 270$	$65 - 165$
n (cm^{-3})	$8 - 3.5 \times 10^3$	$0.4 - 10$
ϵ_e	$0.01 - 0.45$	$0.07 - 0.5$
ϵ_B	$5 \times 10^{-6} - 0.4$	$1 \times 10^{-3} - 0.45$
$M_{\mathrm{ej,iso}}$ (10^{-4} M$_\odot$)	$1 - 40$	$2 - 18$
E_{ST} (10^{50} erg)	$9 - 56$	$15 - 38$
$E_K(t_{\mathrm{dec}})$ (10^{51} erg)	4	30

Note. — Physical parameters of GRBs 980703 and 970508 derived from the non-relativistic evolution of their blastwaves. The range of allowed radii, and hence physical parameters, is determined by the condition $(E_e + E_B) \leq E_{\mathrm{ST}}/2$. The last entry in the table, $E_K(t_{\mathrm{dec}})$, is the total kinetic energy at the deceleration time, $t_{\mathrm{dec}} \approx 90$ s, including synchrotron radiative losses (§5.5).

CHAPTER 6

A Common Origin for Cosmic Explosions Inferred from Calorimetry of GRB 030329[†]

E. Berger[a], S. R. Kulkarni[a], G. Pooley[b], D. A. Frail[c], V. McIntyre[d], R. M. Wark[e], R. Sari[f], A. M. Soderberg[a], D. W. Fox[a], S. Yost[g], P. A. Price[h]

[a]Department of Astronomy, 105-24 California Institute of Technology, Pasadena, CA 91125, USA

[b]Mullard Radio Astronomy Observatory, Cavendish Lab., Madingley Road, Cambridge CB3 0HE, UK

[c]National Radio Astronomy Observatory, P. O. Box 0, Socorro, NM 87801

[d]Australia Telescope National Facility, CSIRO, P.O. Box 76, Epping, NSW 1710, Australia

[e]Australia Telescope National Facility, CSIRO, Locked Bag 194, Narrabri NSW 2390, Australia

[f]Theoretical Astrophysics 130-33, California Institute of Technology, Pasadena, CA 91125, USA

[g]Space Radiation Laboratory 220-47, California Institute of Technology, Pasadena, CA 91125, USA

[h]RSAA, ANU, Mt. Stromlo Observatory, via Cotter Rd, Weston Creek, ACT, 2611, Australia

Abstract

Past studies (Frail et al. 2001; Berger et al. 2003a; Bloom et al. 2003b) suggest that long-duration γ-ray bursts (GRBs) have a standard energy of $E_\gamma \sim 10^{51}$ erg in ultra-relativistic ejecta when corrected for asymmetry ("jets"). However, recently (Berger et al. 2003a; Bloom et al. 2003b) a group of sub-energetic bursts, including the peculiar GRB 980425 associated (Galama et al. 1998c) with SN 1998bw ($E_\gamma \approx 10^{48}$ erg), has been identified. Here we report radio observations of GRB 030329, the nearest burst to date, which allow us to undertake calorimetry of the explosion. Our observations require a two-component explosion: a narrow ($5°$) ultra-relativistic component responsible for the γ-rays and early afterglow, and a wide, mildly relativistic component responsible for the radio and optical afterglow beyond 1.5 days. While the γ-rays are energetically minor, the total energy release, dominated by the wide component, is similar (Frail et al. 2001; Berger et al. 2003a; Bloom et al. 2003b; Panaitescu & Kumar 2002) to that of other GRBs. Given the firm link (Stanek et al. 2003; Hjorth & et al. 2003) of GRB 030329 with SN 2003dh our result suggests a common origin for cosmic explosions in which, for reasons not understood, the energy in the highest velocity ejecta is highly variable.

[†] A version of this chapter was published in *Nature*, vol. 426, 154–157, (2003).

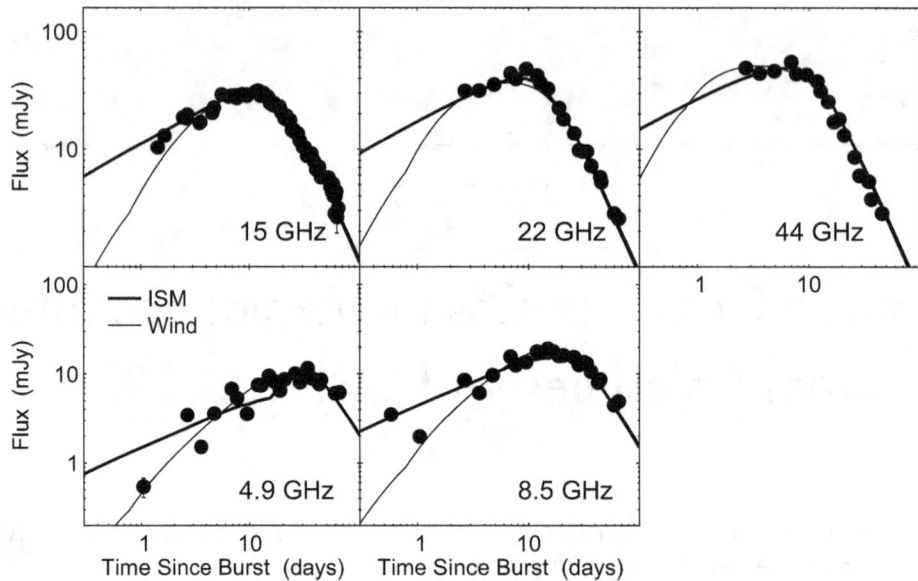

Figure 6.1: Radio light curves of the afterglow of GRB 030329. All measurements include 1σ error bars which in most cases are smaller than the symbols. The data are summarized in Tables 6.1 and 6.2. The solid lines are models of synchrotron emission from collimated relativistic ejecta expanding into uniform (thick) and wind (thin) circumburst media.

SECTION 6.1

Radio Observations of GRB 030329

We initiated observations of the nearby GRB 030329 ($z = 0.1685$) in the centimeter band with the Very Large Array (VLA) approximately 13.8 hours after the burst, on March 30.06 UT. A single 7-hour observation was obtained with the Australia Telescope Compact Array (ATCA) on Mar. 30.53 UT. Radio observations at 15.3 GHz made with the Ryle Telescope at Cambridge (UK). The log of the observations and the resulting light curves are displayed in Tables 6.1 and 6.2 and Figure 6.1.

In the initial observation we detect a point source at right ascension α(J2000)=$10^h 44^m 49.95^s$, and declination δ(J2000)=$21°31'17.38''$, with an uncertainty of about 0.1 arcsec in each coordinate, consistent with the position of the optical counterpart.

In all VLA observations we used the standard continuum mode with 2×50 MHz bands. At 22.5 and 43.3 GHz we used referenced pointing scans to correct for the systematic $10 - 20$ arcsec pointing errors of the VLA antennas. We used the extra-galactic sources 3C 147 (J0542+498) and 3C 286 (J1331+305) for flux calibration, while the phase was monitored using J1111+199 at 1.43 GHz and J1051+213 at all other frequencies. The ATCA observations were performed at 4.80, 6.21, 8.26, and 9.02 GHz with a bandwidth of 64 MHz in each frequency. The phase was monitored using J1049+215, while the flux was calibrated using J1934-638. The data were reduced and analyzed using the Astronomical Image Processing System (VLA) and the Multichannel Image Reconstruction, Image Analysis and Display package (ATCA). The flux density and uncertainty were measured from the resulting maps by fitting a Gaussian model to the afterglow. In addition to the rms noise in each measurement we estimate a systematic uncertainty of about 2% due to uncertainty in the absolute flux calibration.

All observations with the Ryle telescope were made by interleaving 15 minutes scans of GRB 030329 with 2.5 minutes scans of the phase calibrator J1051+2119. The absolute flux scale was calibrated using 3C 48 and 3C 286. We used 5 antennas providing 10 baselines in the range $35 - 140$ m. Since the position of the source is well known the in-phase component of the vector sum of the 10 baselines was

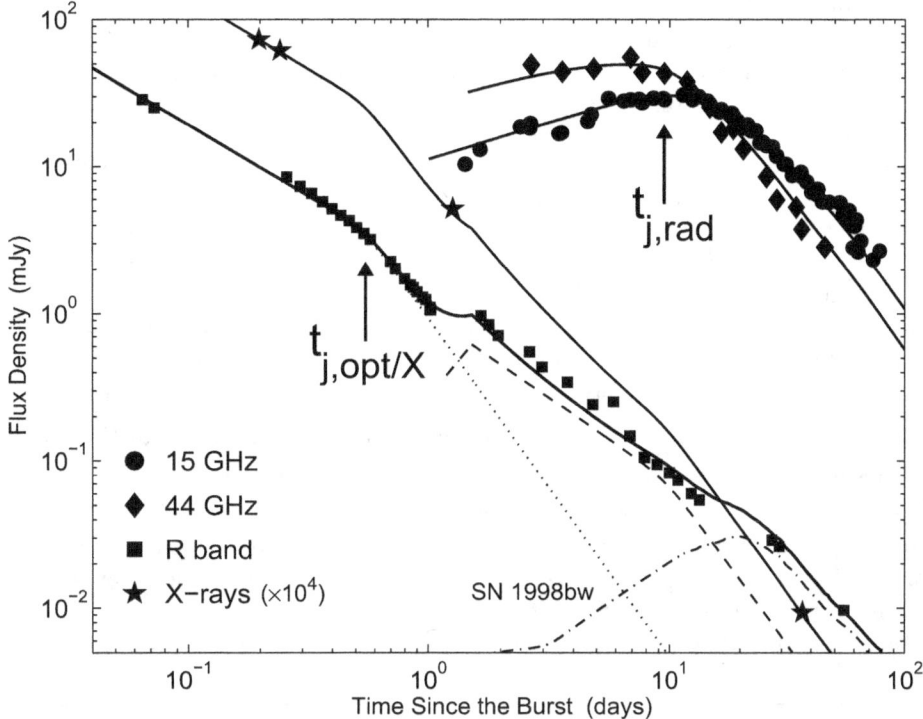

Figure 6.2: Radio to X-ray light curves of the afterglow of GRB 030329. The optical data, from Price et al. (2003) and the GRB Coordinates Network (Henden et al. 2003; Ibrahimov et al. 2003; Testa et al. 2003), have been corrected for Galactic extinction, $A_R = 0.067$ mag. The dotted line is the model proposed by Price et al. (2003) for the early optical emission, with $t_{j,\mathrm{opt}} \approx 0.55$ d. The dashed line is an extrapolation of our uniform density model to the optical R-band. The model in the X-ray band is based on the measured (Tiengo et al. 2003) optical to X-ray spectral slope and an extrapolation of our uniform density model. The sharp increase in the optical flux at $t \lesssim 1.5$ d is due to the deceleration of the slower second jet component. Finally, the dot-dashed line is the optical emission from SN 1998bw at the redshift of GRB 030329, $z = 0.1685$, used as a proxy for SN 2003dh (Stanek et al. 2003).

used as an unbiased estimate of the flux density. The typical rms fluctuation on the signal in a 32-s integration period is approximately 6 mJy. We also add a systematic uncertainty of about 2% due to uncertainty in the absolute flux calibration.

The afterglow was also observed extensively in the millimetre (100 GHz) and sub-millimetre (250 GHz) bands (Sheth et al. 2003). While this is the brightest radio afterglow detected to date, the low redshift results in a peak luminosity, $L_{\nu,p}(8.5\,\mathrm{GHz}) \approx 1.8 \times 10^{31}$ erg s^{-1} Hz^{-1}, typical (Frail et al. 2003a) of other long-duration GRBs.

SECTION 6.2

Broad-band Afterglow Models

The observed rapid decline, $F_\nu \propto t^{-1.9}$ at $t \gtrsim 10$ d and the decrease in peak flux at $\nu \lesssim 22.5$ GHz (Figure 6.1) are the hallmarks of a collimated explosion. In this framework (Sari et al. 1999), the sharp decline (or "jet break") occurs at the time, t_j, when $\Gamma(t_j) \sim \theta_j^{-1}$ due to relativistic abberation ("beaming") and rapid side-ways expansion; here Γ is the bulk Lorentz factor and θ_j is the opening angle of the jet.

We model the afterglow emission (e.g., Berger et al. 2000; Panaitescu & Kumar 2002) from 4.9 to 250

GHz assuming a uniform (Sari et al. 1999) as well as a "wind" (Chevalier & Li 2000) (particle density profile, $\rho \propto r^{-2}$, where r is the distance from the source) circumburst medium. We find $\chi_r^2 = 31.3$ and 39.8 (164 degrees of freedom) for the uniform density and wind models, respectively; these include a 2% systematic error added in quadrature to each measurement. The large values of χ_r^2 are dominated by interstellar scintillation (ISS) at $\nu \lesssim 15$ GHz and mild deviations from the expected smooth behavior at the high frequencies. Comparing the data and models, we find rms flux modulations of 0.25 at 4.9 GHz, 0.15 at 8.5 GHz, and 0.08 at 15 GHz, as well as a drop by a factor of three in the level of modulation from ~ 3 to 40 days. These properties are expected in weak ISS as the fireball expands on the sky. The inferred source size of about 20 μas (i.e., $\sim 2 \times 10^{17}$ cm) at $t \sim 15$ days is in close agreement with theoretical expectations (Galama et al. 2003).

In the uniform density model the jet break occurs at $t \approx 10$ d corresponding to an opening angle, $\theta_j \approx 0.3$ (17°). From the derived synchrotron parameters (at $t = t_j$): $\nu_a \approx 19$ GHz, $\nu_m \approx 43$ GHz, $F_{\nu,m} \approx 96$ mJy we find an isotropic kinetic energy, $E_{K,\mathrm{iso}} \approx 5.6 \times 10^{51} \nu_{c,13}^{1/4}$ erg, a circumburst density $n = 1.8 \nu_{c,13}^{3/4}$ cm^{-3}, and the fractions of energy in the relativistic electrons and magnetic field of $0.16\nu_{c,13}^{1/4}$ and $0.10\nu_{c,13}^{-5/4}$, respectively; here $\nu_c = 10^{13}\nu_{c,13}$ is the synchrotron cooling frequency, and a constraint on Inverse Compton cooling as advocated by Sari & Esin (Sari & Esin 2001) indicates $\nu_{c,13} \lesssim 1$. The beaming-corrected kinetic energy is $E_K \approx 2.5 \times 10^{50} \nu_{c,13}^{1/4}$ erg, typical of other well-studied long-duration GRBs (Panaitescu & Kumar 2002). The parameters derived from the wind model are consistent with those from the uniform density model to within 10%.

Thus, neither model is strongly preferred, but $t_{j,\mathrm{rad}} \approx 9.8$ d is required (Figure 6.1).

SECTION 6.3

A Two-Component Jet

Using the inferred particle density of $n \approx 1.8$ cm^{-3} and assuming a γ-ray efficiency, $\epsilon_\gamma = 0.2$ (Bloom et al. 2003b) we infer $\theta_{j,\mathrm{rad}} \sim 0.3$ rad, or 17°. The kinetic energy in the explosion corrected for collimation is $E_K = f_b E_{K,\mathrm{iso}} \approx 2.5 \times 10^{50}$ erg, where $f_b = [1 - \cos(\theta_j)]$ is the beaming fraction and $E_{K,\mathrm{iso}}$ is the isotropic equivalent kinetic energy. This value is comparable to that inferred from modeling of other afterglows (Panaitescu & Kumar 2002).

In contrast to the above discussion, Price et al. (2003) note a sharp break in the optical afterglow at $t = 0.55$ d (Figure 6.2). The X-ray flux (Tiengo et al. 2003) tracks the optical afterglow for the first day, with a break consistent with that seen in the optical. Thus the break at 0.55 d is not due to a change in the ambient density since for typical parameters (Kumar 2000; Freedman & Waxman 2001) the X-ray emission is not sensitive to density. However, unlike the optical emission the X-ray flux at later times continues to decrease monotonically. Thus we conclude that there are two emitting components: one responsible for the early optical and X-ray emission and the other responsible for the optical emission beyond 1.5 days.

The first component, given the characteristic t^{-2} decay for both the X-ray and optical emission, is reasonably modeled by a jet. For the parameters used above (n, ϵ_γ) the opening angle is 0.09 rad or 5°.

The resurgence in the optical emission at 1.5 d requires a second component. An increase in the ambient density cannot explain this resurgence since the predicted decrease in radio luminosity, arising from the increase in synchrotron self-absorption, is not observed (Figure 6.1). An increase in the energy of the first component, for example by successive shells with lower Lorentz factors as advocated by Granot et al. (Granot et al. 2003), is ruled out by the lack (Sheth et al. 2003) of strong radio or millimetric emission expected (Sari & Mészáros 2000) from reverse shocks.

Thus, by a process of elimination, we are led to a two-component explosion model in which the first component (a narrow jet, 5°) with initially larger Γ is responsible for the γ-ray burst and the early optical and X-ray afterglow including the break at 0.55 d, while the second component (a wider jet, 17°) powers the radio afterglow and late optical emission (Figure 6.2). The break due to the second

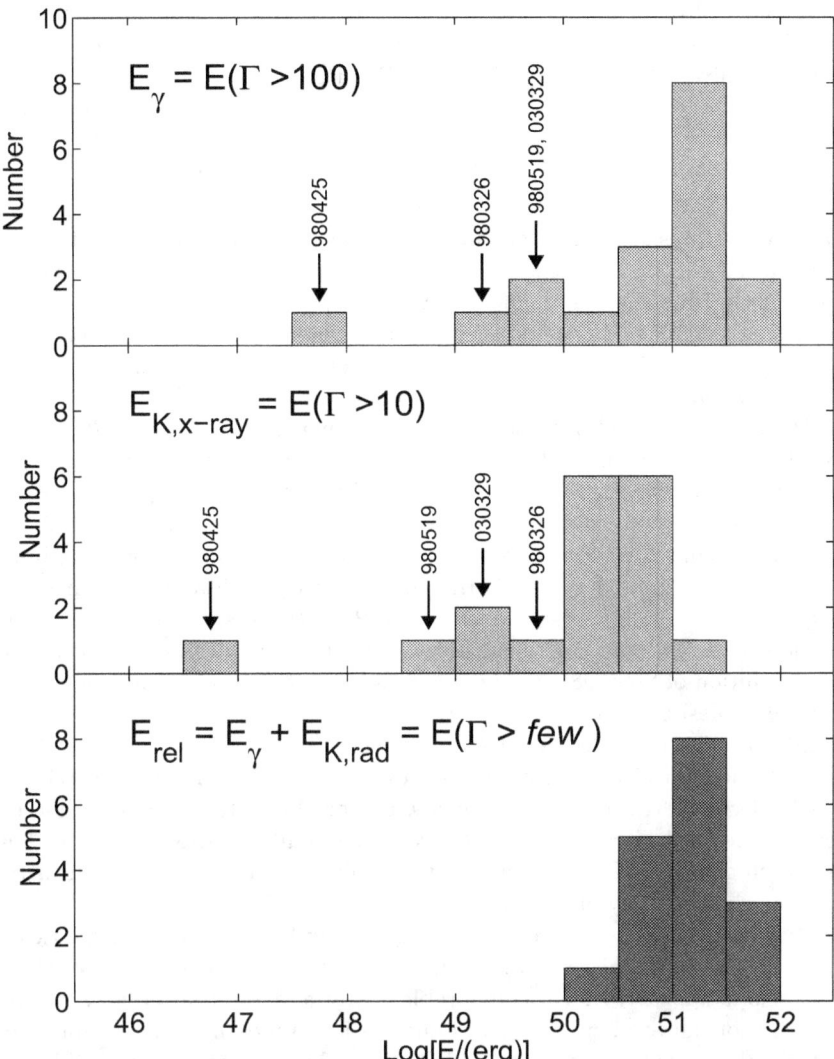

Figure 6.3: Histograms of various energies measured for GRBs: the beaming-corrected γ-ray energy, E_γ, the kinetic energy inferred from X-rays at $t = 10$ hr, $E_{K,X}$, and the total relativistic energy, $E_{\rm rel} = E_\gamma + E_K$, where E_K is the beaming-corrected kinetic energy inferred (Li & Chevalier 1999; Panaitescu & Kumar 2002) from the broad-band afterglow. The significantly wider dispersion in E_γ and $E_{K,X}$ as compared to the total explosive yield indicates that engines in cosmic explosions produce approximately the same quantity of energy (thus pointing to a common origin), but the quality of these engines, as indicated by ultra-relativistic output, varies widely.

component is readily seen in the radio afterglow, but is masked by SN 2003dh in the optical bands, thus requiring careful subtraction (Figure 6.2). Such a two-component jet finds a natural explanation in the collapsar model (MacFadyen et al. 2001).

The beaming-corrected γ-ray energy, emitted by the narrow jet, is only $E_\gamma \approx 5 \times 10^{49}$ erg, significantly lower than the strong clustering (Bloom et al. 2003b) around 1.3×10^{51} erg seen in most bursts. Similarly, the beaming-corrected X-ray luminosity (Tiengo et al. 2003) at $t = 10$ hours, a proxy for the kinetic energy of the afterglow on that timescale, is $L_{X,10} \approx 3 \times 10^{43}$ erg s^{-1}, a factor of ten below the tightly clustered values (Berger et al. 2003a) for most other bursts. However, the second component, which is mildly relativistic (as determined by the lower energy peak of its spectrum), carries the bulk of the energy, as indicated by our modeling of the radio emission. We note that our model, with the energy in the lower Lorentz factor component dominating over the narrow ultra-relativistic component, is not consistent with "universal standard jet" model (Rossi et al. 2002).

SECTION 6.4

A Common Origin for Cosmic Explosions

The afterglow calorimetry presented here has important ramifications for our understanding of GRB engines. Recently, we have come to recognize a sub-class of cosmological GRBs marked by rapidly fading afterglows at early time (i.e., similar to GRB 030329). These events are sub-energetic (Berger et al. 2003a; Bloom et al. 2003b) in E_γ and early X-ray afterglow luminosity. However, as demonstrated by our calorimetry of GRB 030329, such bursts may have total explosive yields similar to other GRBs.

In Figure 6.3 we plot E_γ (Bloom et al. 2003b), the kinetic energy inferred from X-rays at $t = 10$ hr (Berger et al. 2003a), $E_{K,X}$, and the total relativistic energy, $E_{\rm rel} = E_\gamma + E_K$, where E_K is the beaming-corrected kinetic energy inferred (Li & Chevalier 1999; Panaitescu & Kumar 2002) from the broad-band afterglow. The energy in X-rays is determined using $E_{K,X} = L_X t / \epsilon_e (\alpha_X - 1)$, with $t = 10$ hr, $\epsilon_e = 0.1$, and $\alpha_X = 1.3$ is the median decay rate in the X-ray band. For GRB 980519 we find that the evolution of the radio emission requires a much wider jet, $\theta_j \sim 0.3$, than what is inferred from the optical, $\theta_j \sim 0.05$; here we assume $z = 1$. We therefore infer $E_K \sim 2 \times 10^{50}$ erg from the radio data compared to $E_\gamma \approx 4 \times 10^{49}$ erg. The γ-ray energy of GRB 980425 is an upper limit since the degree of collimation is not known. For the kinetic energy we use the value derived by Li & Chevalier (1999) based on the radio evolution of SN 1998bw. There is a significantly wider dispersion in E_γ and $E_{K,X}$ as compared to the total explosive yield.

This leads to the following conclusions. First, radio calorimetry, which is sensitive to all ejecta with $\Gamma \gtrsim few$, shows that the explosive yield of the nearest "classical" event, GRB 030329, is dominated by mildly relativistic ejecta. Ultra-relativistic ejecta which produced the γ-ray emission is energetically unimportant. Second, the total energy yield of GRB 030329 is similar to those estimated for other bursts. Along these lines, the enigmatic GRB 980425 associated (Galama et al. 1998c) with the nearby supernova SN 1998bw also has negligible γ-ray emission, $E_{\gamma,\rm iso} \approx 8 \times 10^{47}$ erg; however, radio calorimetry (Li & Chevalier 1999) shows that even this extreme event had a similar explosive energy yield (Figure 6.3). The newly recognized class of cosmic explosions, the X-ray Flashes (Heise et al. 2003), exhibits little or no γ-ray emission but appear to have comparable X-ray and radio afterglows to those of GRBs. Thus, the commonality of the total energy yield indicates a common origin, but apparantly the ultra-relativistic output is highly variable. Unraveling what physical parameter is responsible for the variation in the "purity" (ultra-relativistic output) of the engine appears to be the next frontier in the field of cosmic explosions.

GRB research at Caltech is supported in part by funds from NSF and NASA. We are, as always, indebted to Scott Barthelmy and the GCN. The VLA is operated by the National Radio Astronomy Observatory, a facility of the National Science Foundation operated under cooperative agreement by Associated Universities, Inc. The Australia Telescope is funded by the Commonwealth of Australia for operations as a National Facility managed by CSIRO. The Ryle Telescope is supported by PPARC.

Table 6.1.   Very Large Array Radio Observations of GRB 030329

Epoch UT	Δt (days)	$F_{1.43}$ (mJy)	$F_{4.86}$ (mJy)	$F_{8.46}$ (mJy)	$F_{15.0}$ (mJy)	$F_{22.5}$ (mJy)	$F_{43.3}$ (mJy)
Mar 30.06	0.58	—	—	3.50 ± 0.06	—	—	—
Mar 30.53	1.05	—	—	1.98 ± 0.17	—	—	—
Apr 1.13	2.65	< 0.21	0.54 ± 0.13	8.50 ± 0.05	19.68 ± 0.14	30.40 ± 0.06	46.63 ± 0.18
Apr 2.05	3.57	< 0.30	1.51 ± 0.05	6.11 ± 0.04	16.98 ± 0.19	31.59 ± 0.14	44.17 ± 0.35
Apr 3.21	4.76	< 0.36	3.58 ± 0.04	9.68 ± 0.03	22.59 ± 0.12	35.57 ± 0.09	46.32 ± 0.23
Apr 5.37	6.89	< 0.40	6.77 ± 0.08	15.56 ± 0.06	28.58 ± 0.20	44.09 ± 0.15	55.33 ± 0.43
Apr 6.16	7.68	< 0.25	5.34 ± 0.10	12.55 ± 0.21	27.26 ± 0.21	39.68 ± 0.20	43.81 ± 1.00
Apr 7.97	9.49	< 0.68	3.55 ± 0.11	13.58 ± 0.09	28.50 ± 0.23	48.16 ± 0.23	43.06 ± 1.33
Apr 10.38	11.90	< 0.58	7.51 ± 0.08	17.70 ± 0.05	31.40 ± 0.25	42.50 ± 0.14	37.86 ± 0.46
Apr 11.17	12.69	—	7.42 ± 0.09	17.28 ± 0.10	29.60 ± 0.29	36.84 ± 0.16	31.26 ± 0.51
Apr 13.35	14.87	—	9.49 ± 0.13	19.15 ± 0.08	26.78 ± 0.33	32.69 ± 0.13	25.44 ± 0.51
Apr 15.14	16.66	—	8.21 ± 0.08	17.77 ± 0.10	24.50 ± 0.31	—	17.10 ± 0.71
Apr 17.20	18.72	< 0.63	6.50 ± 0.11	15.92 ± 0.07	22.02 ± 0.25	22.41 ± 0.08	18.07 ± 0.28
Apr 19.06	20.58	—	8.66 ± 0.10	16.08 ± 0.06	18.35 ± 0.24	18.03 ± 0.11	13.15 ± 0.29
Apr 24.18	25.70	—	10.04 ± 0.08	15.34 ± 0.06	13.93 ± 0.26	13.63 ± 0.13	8.54 ± 0.48
Apr 26.92	28.44	< 0.58	8.05 ± 0.08	12.67 ± 0.09	11.82 ± 0.26	9.75 ± 0.23	5.95 ± 0.62
Apr 28.96	30.48	—	9.80 ± 0.09	—	10.40 ± 0.33	9.53 ± 0.21	—
Apr 29.99	31.51	—	11.62 ± 0.08	13.55 ± 0.07	—	—	—
May 2.06	33.58	—	—	13.10 ± 0.06	—	9.52 ± 0.14	—
May 3.07	34.59	—	8.90 ± 0.08	—	—	—	5.30 ± 0.32
May 5.00	36.52	—	7.72 ± 0.13	10.64 ± 0.06	8.58 ± 0.17	7.20 ± 0.09	3.75 ± 0.26
May 11.03	42.55	—	8.57 ± 0.09	8.04 ± 0.08	7.03 ± 0.19	5.75 ± 0.10	—
May 13.03	44.55	—	—	8.68 ± 0.08	5.77 ± 0.22	5.23 ± 0.17	—
May 14.00	45.52	—	6.08 ± 0.10	—	—	2.84 ± 0.20	2.84 ± 0.23
May 28.03	59.55	—	6.20 ± 0.08	4.48 ± 0.09	2.82 ± 0.21	2.56 ± 0.12	—
June 4.01	66.53	1.94 ± 0.06		4.93 ± 0.06			

Note. — The columns are (left to right), (1) Epoch of observation, (2) time since the burst, and (3-8) measured flux densities at 1.43 through 43.3 GHz.

Table 6.2.   Ryle Telescope Radio Observations of GRB 030329

Epoch UT	Δt (days)	$F_{15.3}$ (mJy)	Epoch UT	Δt (days)	$F_{15.3}$ (mJy)
Mar 30.91	1.43	10.38 ± 0.28	Apr 21.72	23.24	17.63 ± 0.29
Mar 31.12	1.64	13.05 ± 0.28	Apr 22.66	24.18	14.51 ± 0.49
Mar 31.91	2.43	18.66 ± 0.28	Apr 23.33	24.85	14.62 ± 0.49
Apr 1.12	2.64	18.29 ± 0.28	Apr 25.81	27.33	13.60 ± 0.65
Apr 1.98	3.50	16.75 ± 0.27	Apr 26.82	28.34	11.78 ± 0.52
Apr 3.07	4.59	20.36 ± 0.45	Apr 29.82	31.34	10.35 ± 0.49
Apr 4.09	5.61	29.13 ± 0.52	May 1.63	33.15	8.73 ± 0.52
Apr 4.97	6.49	27.97 ± 0.26	May 4.80	36.32	9.15 ± 0.50
Apr 5.97	7.49	28.69 ± 0.26	May 6.83	38.35	7.87 ± 0.50
Apr 7.06	8.58	29.29 ± 0.49	May 8.73	40.25	6.70 ± 0.50
Apr 7.89	9.41	29.15 ± 0.44	May 10.76	42.28	6.49 ± 0.50
Apr 9.89	11.41	30.78 ± 0.51	May 15.76	47.28	5.74 ± 0.50
Apr 11.05	12.57	28.52 ± 0.51	May 20.70	52.22	5.69 ± 0.53
Apr 11.88	13.40	29.92 ± 0.44	May 22.76	54.28	4.78 ± 0.78
Apr 13.05	14.57	27.90 ± 0.44	May 24.76	56.28	4.31 ± 0.55
Apr 13.87	15.39	24.74 ± 0.44	May 25.56	57.08	5.04 ± 0.84
Apr 14.82	16.34	23.60 ± 0.32	May 26.75	58.27	3.99 ± 0.63
Apr 16.96	18.48	23.06 ± 0.24	May 28.76	60.28	3.96 ± 0.58
Apr 17.92	19.44	20.51 ± 0.24	May 29.82	61.34	4.35 ± 0.50
Apr 19.95	21.47	19.27 ± 0.38	May 30.76	62.28	2.65 ± 0.72
Apr 20.72	22.24	17.53 ± 0.33	June 2.54	64.06	3.13 ± 0.76

Note. — The columns are (left to right), (1) Epoch of observation, (2) time since the burst, and (3) measured flux densitat 15.3 GHz.

Part II

The Search for Engine-Driven Supernovae

CHAPTER 7

The Radio Evolution of the Ordinary Type Ic SN 2002ap[†]

E. Berger[a], S. R. Kulkarni[a], & R. A. Chevalier[b]

[a]Department of Astronomy, 105-24 California Institute of Technology, Pasadena, CA 91125, USA

[b]Department of Astronomy, University of Virginia, P.O. Box 3818, Charlottesville, VA 22903-0818

Abstract

We report the discovery and monitoring of radio emission from the Type Ic SN 2002ap ranging in frequency from 1.43 to 22.5 GHz, and in time from 4 to 50 days after the SN explosion. As in most other radio SNe, the radio spectrum of SN 2002ap shows evidence for absorption at low frequencies, usually attributed to synchrotron self-absorption or free-free absorption. While it is difficult to discriminate between these two processes based on a goodness-of-fit, the *unabsorbed* emission in the free-free model requires an unreasonably large ejecta energy. Therefore, on physical grounds we favor the synchrotron self-absorption (SSA) model. In the SSA framework, at about day 2, the shock speed is $\approx 0.3c$, the energy in relativistic electrons and magnetic fields is $\approx 1.5 \times 10^{45}$ erg and the inferred progenitor mass loss rate is $\approx 5 \times 10^{-7}$ $M_\odot$ yr^{-1} (assuming a 10^3 km sec^{-1} wind). These properties are consistent with a model in which the outer, high velocity supernova ejecta interact with the progenitor wind. The amount of relativistic ejecta in this model is small, so that the presence of broad lines in the spectrum of a Type Ib/c supernova, as observed in SN 2002ap, is not a reliable indicator of relativistic ejecta and hence γ-ray emission.

SECTION 7.1
Introduction

Type Ib/c supernovae (SNe) enjoyed a broadening in interest over the last few years since their compact progenitors (Helium or Carbon stars) are ideal for detecting the signatures of a central engine. Such an engine is expected in the collapsar model (Woosley 1993; MacFadyen et al. 2001), the currently popular model for long-duration γ-ray bursts (GRBs). In this model, the engine (a rotating and accreting black hole) provides the dominant source of explosive power. The absence of an extensive Hydrogen envelope in the progenitor star may allow the jets from the central engine to propagate to the surface and subsequently power bursts of γ-rays.

[†] A version of this chapter was published in *The Astrophysical Journal*, vol. 577, L5–L8, (2002).

Separately, the Type Ic SN 1998bw (Galama et al. 1998a) found in the localization region of GRB 980425 (Pian et al. 2000) ignited interest in "hypernovae"[1] . SN 1998bw is peculiar for three reasons: (i) broad photospheric absorption lines (Iwamoto et al. 1998; Woosley et al. 1999), (ii) a large kinetic energy release, $E_{k,51} \sim 30$ ($E_k = 10^{51} E_{k,51}$ erg is the SN energy), inferred from the optical data and (iii) bright radio emission at early time. Robust equipartition arguments led to an inferred energy of $E_\Gamma \gtrsim 10^{49}$ erg in ejecta with relativistic velocities, $\Gamma \sim$ few (Kulkarni et al. 1998). No other SN has shown hints of such an abundance of relativistic ejecta. Tan et al. (2001) explain the relativistic ejecta as resulting from an energetic shock as it speeds up the steep density gradient of the progenitor. The γ-ray and radio emission would then arise in the forward shock.

From the perspective of a GRB–SN connection, what matters most is the presence of relativistic ejecta. γ-ray emission traces ultra-relativistic ejecta, but as was dramatically demonstrated by SN 1998bw, the radio serves as an equally good proxy, with the added advantage that the emission is not beamed. Given this, we began a systematic program of investigating at radio wavelengths all Ib/c SNe with features similar to SN 1998bw: a hypernova or broad optical lines.

Y. Hirose discovered SN 2002ap in M74 (distance, $d \sim 7.3$ Mpc; Smartt et al. 2002) on 2002, Jan. 29.40 UT (see Nakano et al. 2002). Mazzali et al. (2002) inferred an explosion date of 2002, Jan. 28 ± 0.5 UT. Attracted by the broad spectral features (e.g., Kinugasa et al. 2002; Meikle et al. 2002) we began observing the SN at the Very Large Array (VLA[2]).

SECTION 7.2

Observations

We observed SN 2002ap starting on 2002, February 1.03 UT, and detected a radio source coincident with the optical position at α(J2000)= $01^h 36^m 23.92^s$, δ(J2000)=+15°45'12.87", with a 1-σ uncertainty of 0.1 arcsec in each coordinate (Berger et al. 2002c). A log of the observations and the resulting light curves can be found in Table 7.1 and Figure 7.1, respectively.

7.2.1   The Radio Spectrum of SN 2002ap

The peak radio luminosity of SN 2002ap, $L_p(5\,\text{GHz}) \sim 3 \times 10^{25}$ erg sec^{-1} Hz^{-1}, is a factor of 20 lower than the typical Ib/c SNe (Weiler et al. 1998), and $\sim 3 \times 10^3$ times lower than SN 1998bw (Kulkarni et al. 1998). The time at which the radio emission peaks at 5 GHz is $t_p \sim 3$ day, which may be compared with 10 days for SN 1998bw , and 10–30 days for the typical Ib/c SNe (Weiler et al. 1998; Chevalier 1998).

The spectral index between 1.43 and 4.86 GHz, $\beta_{1.43}^{4.86}$, ($F_\nu \propto \nu^\beta$) changes from ~ 0.5 before day 6, to ~ -0.3 at $t \approx 15$ days, while $\beta_{4.86}^{8.46}$ holds steady at a value of ≈ -0.9. This indicates that the spectral peak, ν_p, is initially located between 1.43 and 4.86 GHz, and decreases with time. This peak could be due to synchrotron self-absorption (SSA) or (predominantly) free-free absorption (FFA) arising in the circumstellar medium (CSM). Regardless of the dominant source of opacity, the emission for frequencies $\nu > \nu_p$ is from optically-thin synchrotron emission.

Massive stars lose matter via strong stellar winds throughout their life and as a result their CSM is inhomogeneous with density, $\rho(r) \propto \dot{M}_w v_w^{-1} r^{-2}$. Here, r is the distance from the star, $\dot{M}$ is the rate of mass loss, and v_w is the wind speed, which is comparable to the escape velocity from the star. The progenitors of Type II SNe are giant stars which have low $v_w \sim 10$ km s^{-1}. Consequently the CSM is dense and this explains why the FFA model has provided good fits to Type II SNe (e.g., Weiler et al. 1998).

---

[1] There is no accepted definition for a hypernova. Here we use the term to mean a supernova with an explosion energy significantly larger than 10^{51} erg.

[2] The VLA is operated by the National Radio Astronomy Observatory, a facility of the National Science Foundation operated under cooperative agreement by Associated Universities, Inc.

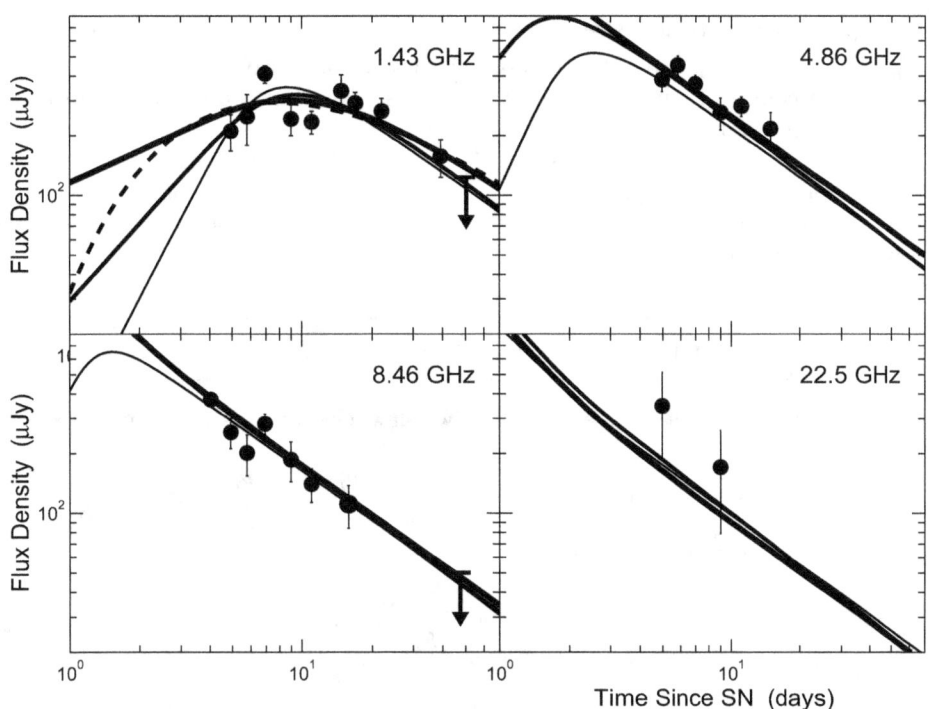

Figure 7.1: Radio light curves of SN 2002ap. The thick solid lines are our three synchrotron self-absorption models described in §7.3, with $\tau_\nu \propto t^{-1.3}$, $\tau_\nu \propto t^{-2.1}$, and $\tau_\nu \propto t^{-3}$ in order of decreasing thickness. The dashed line is the model-fit based on free-free absorption (§7.4). At 4.86, 8.46, and 22.5 GHz, the SSA and FFA models provide the same fit, since the opacity processes do not influence the optically-thin flux. The models diverge in the optically-thick regime, which underlines the importance of rapid, multi-frequency observations.

On the other hand, the progenitors of Type Ib and Ic SNe are compact Helium and Carbon stars which have high escape velocities and therefore fast winds, $\sim 10^3$ km sec^{-1}. Thus, *a priori*, the CSM density is not expected to be high. C98 reviews the modeling of radio emission from Ib/c SNe and concludes that there is little need to invoke free-free absorption. However, synchrotron self absorption is an inescapable source of opacity and must be included in the modeling of Type Ib/c SNe (Chevalier 1998; Kulkarni et al. 1998).

Low-frequency observations provide the simplest way to discriminate between the two models. In the SSA model, the peak frequency is identified with the synchrotron-self absorption frequency, ν_a, and $F_\nu(\nu \lesssim \nu_a) \propto \nu^{5/2}$. In the FFA model, the free-free optical depth is unity at the peak frequency, ν_{ff} and $F_\nu(\nu \lesssim \nu_{ff})$ decreases exponentially. Lacking the requisite discriminatory low frequency data we consider both models.

7.2.2 Robust Constraints

Before performing a detailed analysis, we derive some general constraints using the well-established equipartition arguments (Readhead 1994; Kulkarni et al. 1998). The energy of a synchrotron source with flux density, $F_p(\nu_p)$, can be expressed in terms of the equipartition energy density,

$$\frac{U}{U_{\rm eq}} = \frac{1}{2}\epsilon_B \eta^{11}\left(1 + \frac{\epsilon_e}{\epsilon_B}\eta^{-17}\right), \tag{7.1}$$

where $\eta = \theta_s/\theta_{\rm eq}$, the equipartition size is $\theta_{\rm eq} \approx 120 d_{\rm Mpc}^{-1/17} F_{\rm p,mJy}^{8/17} \nu_{\rm p,GHz}^{(-2\beta-35)/34}$ μas, $U_{eq} = 1.1 \times 10^{56} d_{\rm Mpc}^2 F_{\rm p,mJy}^4 \nu_{\rm p,GHz}^{-7} \theta_{\rm eq,\mu as}^{-6}$ erg, and ϵ_e and ϵ_B are the fractions of energy in the electrons and magnetic fields, respectively. In equipartition $\epsilon_e = \epsilon_B = 1$, and it is clear that a deviation from equipartition would increase the energy significantly.

At about day 7, $F_p(\nu_p = 1.4\,{\rm GHz}) \approx 0.3$ mJy (see Figure 7.1). Thus, $\theta_{\rm eq}(t = 7\,{\rm d}) \approx 40$ μas, or $r \approx 4.5 \times 10^{15}$ cm. The resulting equipartition energy is $E_{\rm eq} \approx 10^{45}$ erg, the magnetic field is $B_{\rm eq} \approx 0.2$ G, and the average velocity of the ejecta is $v_{\rm eq} \approx 0.3c$. We note that any other source of opacity (e.g., free-free absorption) would serve to increase $\theta_{\rm eq}$, $E_{\rm eq}$, and $v_{\rm eq}$.

SECTION 7.3

A Synchrotron Self-Absorption Model

The synchrotron spectrum from a source with a power-law electron distribution, $N(\gamma) \propto \gamma^{-p}$ for $\gamma > \gamma_{\rm min}$ is

$$F_\nu = F_{\nu,0}(\nu/\nu_0)^{5/2}(1 - e^{-\tau_\nu}) \frac{F_3(\nu, \nu_m, p)}{F_3(\nu_0, \nu_m, p)} \frac{F_2(\nu_0, \nu_m, p)}{F_2(\nu, \nu_m, p)}, \tag{7.2}$$

where the optical depth at frequency ν is given by

$$\tau_\nu = \tau_0(\nu/\nu_0)^{-(2+p/2)} \frac{F_2(\nu, \nu_m, p)}{F_2(\nu_0, \nu_m, p)}, \tag{7.3}$$

and

$$F_\ell(\nu, \nu_m, p) = \int_0^{x_m} F(x) x^{(p-\ell)/2} dx; \tag{7.4}$$

see Li & Chevalier (1999). Here $x_m \equiv \nu/\nu_m$ (Rybicki & Lightman 1979), and ν_m is the characteristic synchrotron frequency of electrons with $\gamma = \gamma_{\rm min}$. The subscript zero indicates quantities at a reference frequency which we set to 1 GHz. Finally, ν_a is defined by the equation $\tau_{\nu_a} = 1$.

The evolution of the synchrotron emission depends on a number of parameters. Following Chevalier (1998), we assume that p, ϵ_e, ϵ_B in the post-shock region remain constant with time; here, $\epsilon_e = \epsilon_B = 0.1$. The evolution of the synchrotron spectrum is sensitive to the expansion radius of the forward shock front, $r_s \propto t^m$, which is related to the density structure of the shocked ejecta and that of the CSM. We allow for these hydrodynamic uncertainties by letting $F_{\nu,0} \propto t_d^{\alpha_F}$ and $\tau_0 \propto t_d^{\alpha_\tau}$, where t_d is the time in days since the SN explosion. In the model adopted here, both these indices depend on m and p. It can be shown that the temporal index of the optically thin flux, $\alpha = \alpha_F + \alpha_\tau$. The synchrotron characteristic frequency, ν_m, is particularly useful for inferring the CSM density, and we parametrize it as $\nu_m = \nu_{m,0} t_d^{\alpha_{\nu m}}$ GHz where $\nu_{m,0} = \nu_m$ GHz. For typical values of m, and $\rho(r) \propto r^{-2}$, $\alpha_{\nu_m} \approx -0.9$.

With these scalings and Eqs. 7.2–7.4 we carry out a least-squares fit to the data. Given the lack of early optically-thick data (i.e., 1.43 GHz) it is not surprising that our least-squares analysis allows a broad range of values for α_τ. In Figure 7.1 we plot fits spanning the minimum χ^2: $\alpha_\tau = -1.3, -2.1, -3$ (corresponding to $\chi^2 = 40, 43, 46$, respectively and 21 degrees of freedom). We note that for other Ib/c SNe α_τ range from -2 to -3 (Chevalier 1998; Li & Chevalier 1999)

The fits, in conjunction with Eqs. 13–15 of Li & Chevalier (1999) allow us to trace the evolution of r_s, the total (magnetic+electrons) energy (E), and the electron density (n_e) in the shock (Figure 7.2). We find that for $\alpha_\tau = -1.3$, $r_s \propto t^{0.25}$ i.e., the blastwave decelerates. However, $\alpha_\tau = -3$ provides the expected $r_s \propto t^{0.9}$. Adopting this physically reasonable model, we obtain: $\tau_0(t) = 1.2 \times 10^3 t_d^{-3}$, $F_{\nu,0}(t) = 2.9 t_d^{2.2}$ μJy, and $p = 2$. From Figure 7.2 we note that the early shock velocity is high, $0.3c$, regardless of the choice of α_τ, and close to that derived from the simple equipartition arguments (§7.2.2).

The mass loss rate of the progenitor star is estimated from r_s and n_e, $\dot{M}_w = 8\pi\zeta n_e m_p r^2 v_w \approx 9 \times 10^{-9} \nu_{m,0}^{-0.8}$ M$_\odot$ yr^{-1}, where the compression factor is $\zeta = 1/4$, the nucleon-to-electron ratio is taken to be 2 and $v_w = 10^3$ km sec^{-1}. Knowing B_{eq} and our assumed ϵ_e we find $\nu_m \sim 10^7$ Hz and thus

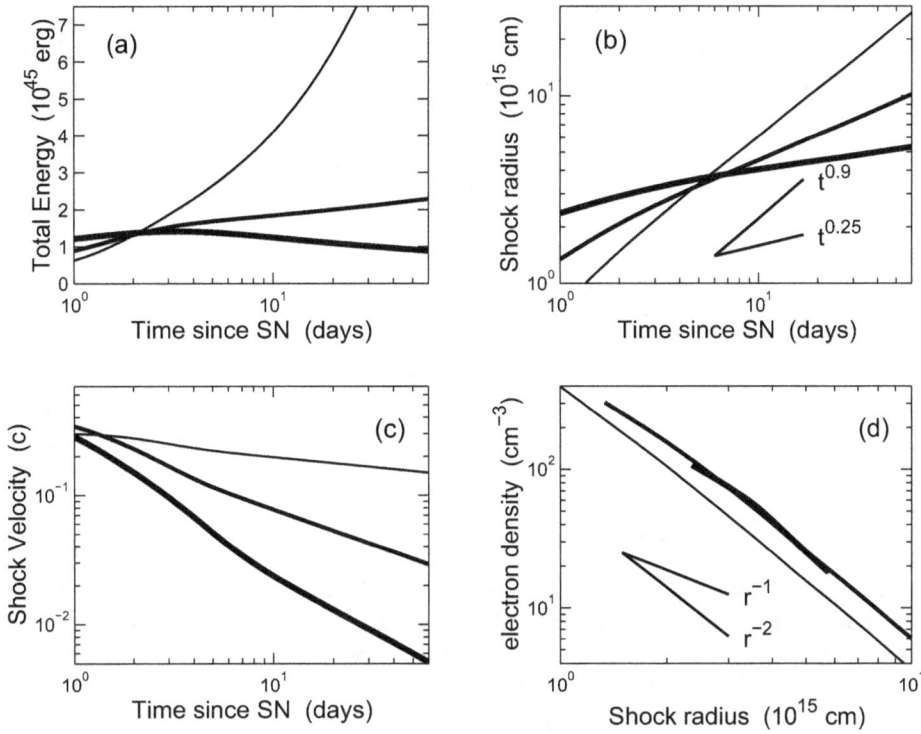

Figure 7.2: Inferred physical parameters based on the synchrotron self-absorption models described in §7.3. The panels are (a) time evolution of the total energy, (b) radius of the radio photosphere, (c) electron density in the shock as a function of radius, and (d) velocity of the shock front as a function of time. Models with $\tau_\nu \propto t^{-1.3}$, $\tau_\nu \propto t^{-2.1}$, and $\tau_\nu \propto t^{-3}$ are shown in order of decreasing thickness. The most likely fit is the one following $r \propto t^{0.9}$ (i.e., the model with $\tau_\nu \propto t^{-3}$).

$\dot{M}_w \approx 5 \times 10^{-7}\, M_\odot\,\mathrm{yr}^{-1}$ – similar to that inferred for SN 1998bw Li & Chevalier (1999). There are two consistency checks. First, with this $\dot{M}_w$, free-free absorption is negligible. Second, the kinetic energy of the swept-up material is 2×10^{46} erg – consistent with our estimate of the equipartition energy and ϵ_e.

7.3.1   The SSA Model in the Context of a Hydrodynamic Model

The results of §7.3 can be tied in to a fairly simple hydrodynamic model. Matzner & McKee (1999) show that for the progenitors of Ib/c SNe (compact stars with radiative envelopes) the ejecta post-explosion density profile can be described by power laws at low and high velocities, separated by a break velocity, $v_{\mathrm{ej,b}} = 5150(E_{k,51}/M_1)^{1/2} \approx 2 \times 10^4$ km sec^{-1}; here the mass of the ejecta is $M_{\mathrm{ej}} = 10 M_1\, M_\odot$. We use $E_{k,51} \approx 4 - 10$ and $M_1 \approx 0.25 - 0.5$ for SN 2002ap (Mazzali et al. 2002). At $v_s \approx 0.3c$, the density profile is given by $\rho_s \approx 3 \times 10^{96} E_{k,51}^{3.59} M_1^{-2.59} t^{-3} v^{-10.18}$ g cm^{-3}. This profile extends until radiative losses become important when the shock front breaks out of the star. Using Eq. 32 of Matzner & McKee (1999) this happens for $v_s \approx 1.5c$ (assuming a typical 1 R$_\odot$ radius for the progenitor star). Thus, the outflow can become relativistic.

Using the self-similar solution of Chevalier (1982) the velocity of the outer shock radius, R, (assuming a $\rho = Ar^{-2}$ CSM) is

$$\frac{R}{t} = 52{,}300\, E_{k,51}^{0.44} M_1^{-0.32} A_*^{-0.12} t_d^{-0.12} \text{ km sec}^{-1}, \tag{7.5}$$

where $A_* = (\dot{M}_w/10^{-5}\ M_\odot\ \text{yr}^{-1})(v_w/10^3\ \text{km sec}^{-1})^{-1}$. The shock velocity, $\dot{R}$, is insensitive to the circumstellar wind density. Thus, we find that the velocities inferred from the radio observations of SN 2002ap can be naturally accounted for by the outer supernova ejecta.

The energy above some velocity V is

$$E(v > V) \approx \int_V^\infty \frac{1}{2}\rho_f v^2 4\pi v^2 t^3 dv = 7.2 \times 10^{44} E_{k,51}^{3.59} M_1^{-2.59} V_5^{-5.18}\ \text{ergs}, \qquad (7.6)$$

where v_5 is the velocity in units of 10^5 km s^{-1}. For the preferred SN 2002ap parameters, $E(v > V) \approx 3.8 \times 10^{48} V_5^{-5.18}$ erg. There is therefore plenty of energy in the high velocity ejecta to account for the observed radio emission, and in fact a kinetic energy, $E_{k,51} = 0.5$, would be sufficient.

Given the over-abundance of $E(v > V_5)$ relative to the energy inferred from the radio emission, we wonder how secure are the estimates of $E_{k,51}$ and M_1 of Mazzali et al. (2002). In particular, these parameters are derived from early optical observations and are subject to asymmetries in the explosion. For SN 1998bw, the asymmetric model of Höflich et al. (1999) yielded $E_{k,51} \sim 2$, an order of magnitude smaller than that obtained from symmetrical models (e.g., Iwamoto et al. 1998).

7.3.2   Interstellar Scattering & Scintillation

Interstellar scattering and scintillation (ISS) is expected for radio SNe (cf. Kulkarni et al. 1998). Indeed, the perceptible random deviations from the model curves (see Figure 7.1), which account for the high χ^2_{min} could arise from ISS.

Using the ISS model of Goodman (1997), and the Galactic free electron model of Taylor & Cordes (1993), we estimate $m_{8.46} \approx 5\%$, $m_{4.86} \approx 10\%$, and $m_{1.43} \approx 40\%$; m_ν is the modulation index (the ratio of the rms to the mean) for each frequency.

We estimate the actual modulation index empirically by adding $m_\nu F_\nu$ in quadrature to each measurement error so that the reduced χ^2_{min} is unity. Here F_ν is the model flux described in §7.3. We find $m_{8.46} \approx 10\%$, $m_{4.86} \approx 20\%$, and $m_{1.43} \approx 30\%$, in good agreement with the theoretical estimates. This provides an independent confirmation of the size, and hence expansion velocity of the ejecta. We note that since the modulation is not severe in any of the bands, the results of §7.3 are quite robust.

SECTION 7.4

A Free-Free Absorption Model

In this model, the spectrum is parametrized as (Chevalier 1984; Weiler et al. 1986):

$$F_\nu = K_1 \nu_5^\beta t_d^\alpha e^{-\tau_\nu}$$
$$\tau_\nu = K_2 \nu_5^{-2.1} t_d^\delta, \qquad (7.7)$$

where $\nu_5 = 5\nu$ GHz. We find an acceptable fit ($\chi^2 = 40$ for 21 degrees of freedom) yielding: $K_1 \approx 2$ mJy, $K_2 \approx 0.4$, $\alpha \approx -0.9$, $\beta \approx -0.9$, and $\delta \approx -0.8$. With these parameters and Eq. 16 of Weiler et al. (1986) we find $\dot{M}_w \approx 5 \times 10^{-5}\ M_\odot\ \text{yr}^{-1}$ for $v_w = 10^3$ km sec^{-1}.

Using our derived parameters, one day after the explosion $\nu_{\text{ff}} \approx 3.2$ GHz, and $F_\nu(\nu_{\text{ff}}) \approx 1.1$ mJy (Figure 7.1). The unabsorbed flux at the peak of the synchrotron spectrum is $F_\nu(\nu_a) \approx 3(\nu_a/3.2\,\text{GHz})^{-0.9}$ mJy (note $\nu_a < \nu_{ff}$ in the FFA model) for which $r_{\text{eq}} \approx 7.5 \times 10^{15}(\nu_a/3.2\,\text{GHz})^{-3/2}$ cm. Thus $v_{\text{eq}} \approx 3c(\nu_a/3.2\,\text{GHz})^{-3/2}$, which corresponds to $\Gamma = 2(\nu_a/3.2\,\text{GHz})^{-1}$ if relativistic effects are taken into account (R. Sari priv. comm.). Alternatively, if we fix the expansion velocity to the optical value, $v_s \approx 3 \times 10^4$ km sec^{-1} (Mazzali et al. 2002), we find a brightness temperature, $T_b \approx 4 \times 10^{13}$ K — clearly in excess of the equipartition temperature, again necessitating a high bulk Lorentz factor, $\Gamma \sim 10^2$.

Thus, even if $\nu_a = \nu_{\text{ff}}$ (in which case free-free opacity would not be necessary in the first place), the FFA model requires truly relativistic ejecta, or alternatively a large departure from equipartition, resulting in $E \approx 7 \times 10^{50} (\nu_a / 3.2\,\mathrm{GHz})^{-9}$ erg (for $v_s \approx 0.5c$ instead of $3c$). Clearly, the energy requirement would increase by many orders of magnitude if $\nu_a \ll \nu_{\mathrm{ff}}$.

SECTION 7.5

Discussion and Conclusions

SN 1998bw exhibited broad photospheric absorption lines and bright radio emission. These two peculiarities made sense in that the simple theory suggested that broad photospheric features are a reliable indicator of relativistic ejecta, a necessary condition for γ-ray emission.

The type Ic SN 2002ap elicited much interest because it too displayed similar broad lines. However, from our radio observations we estimate the energy in relativistic electrons and magnetic fields to be quite modest: $E \approx 2 \times 10^{45}$ ergs in ejecta with a velocity $\approx 0.3c$. Both the energy and speed of the ejecta can be accounted for in the standard hydrodynamical model. Thus, our principal conclusion is that broad photospheric lines are not good predictors of relativistic ejecta.

Moreover, the broad photospheric features led modelers to conclude that SN 2002ap was a hypernova with an explosion energy of $E_{51} \sim 4 - 10$ erg (Mazzali et al. 2002). However, the radio observations suggest that SN 2002ap is not an energetic event. In the same vein, we note that Kawabata et al. (2002) suggest, based on spectro-polarimetric observations, a jet with a speed of $0.23c$ and carrying 2×10^{51} erg. Such a jet, regardless of geometry, would have produced copious radio emission.

We end with two conclusions. First, at least from the perspective of relativisic ejecta, SN 2002ap was an ordinary Ib/c SN. Second, broad photospheric lines appear not to be a good proxy for either an hypernova origin or γ-ray emission.

Dale Frail was involved in various aspects of this project and we are grateful for his help and encouragement. We also wish to acknowledge useful discussions with J. Craig Wheeler. Finally, we thank NSF and NASA for supporting our research.

Table 7.1.   Radio Observations of SN 2002ap

Epoch (UT)	$F_{1.43} \pm \sigma$ (μJy)	$F_{4.86} \pm \sigma$ (μJy)	$F_{8.46} \pm \sigma$ (μJy)	$F_{22.5} \pm \sigma$ (μJy)
2002 Feb 1.03	—	—	374 ± 29	—
2002 Feb 1.93	211 ± 44	384 ± 50	255 ± 44	348 ± 165
2002 Feb 2.79	250 ± 72	453 ± 50	201 ± 47	—
2002 Feb 3.93	410 ± 41	365 ± 38	282 ± 34	—
2002 Feb 5.96	243 ± 43	262 ± 48	186 ± 42	170 ± 91
2002 Feb 8.00	235 ± 31	282 ± 32	140 ± 27	—
2002 Feb 11.76	337 ± 68	217 ± 45	111 ± 27	—
2002 Feb 13.94	292 ± 38	—	—	—
2002 Feb 18.95	266 ± 42	—	—	—
2002 Mar 4.85+11.83	157 ± 34	—	—	—
2002 Mar 18.77+19.97	57 ± 33	—	25 ± 25	—

Note. — The columns are (left to right), (1) UT date of each observation, and flux density and rms noise at (2) 1.43 GHz, (3) 4.86 GHz, (4) 8.46 GHz, and (5) 22.5 GHz. Observations with more than one date have been co-added to increase the signal-to-noise of the detection.

CHAPTER 8

A Radio Survey of Type Ib and Ic Supernovae: Searching for Engine Driven Supernovae[†]

E. BERGER[a], S. R. KULKARNI[a], D. A. FRAIL[b] & A. M. SODERBERG[a]

[a]Department of Astronomy, 105-24 California Institute of Technology, Pasadena, CA 91125, USA

[b]National Radio Astronomy Observatory, P. O. Box 0, Socorro, NM 87801

Abstract

The association of γ-ray bursts (GRBs) and core-collapse supernovae (SNe) of Type Ib and Ic was motivated by the detection of SN 1998bw in the error box of GRB 980425 and the now-secure identification of SN 2003dh in the cosmological GRB 030329. The bright radio emission from SN 1998bw indicated that it possessed some of the unique attributes expected of GRBs, namely, a large reservoir of energy in (mildly) relativistic ejecta and variable energy input. The two popular scenarios for the origin of SN 1998bw are a typical cosmological burst observed off-axis or a member of a new distinct class of supernova explosions (GRB Supernovae, or gSNe). In the former, about 0.5% of local Type Ib/c SNe are expected to be similar to SN 1998bw; for the latter no such constraint exists. Motivated thus, we began a systematic program of radio observations of most reported Type Ib/c SNe accessible to the Very Large Array. Of the 33 SNe observed from late 1999 to the end of 2002 at most one is as bright as SN 1998bw. From this we conclude that the incidence of such events is $\lesssim 3\%$. Furthermore, analysis of the radio emission indicates that none of the observed SNe exhibit clear engine signatures. Finally, a comparison of the SN radio emission to that of GRB afterglows indicates that none of the SNe could have resulted from a typical GRB, independent of the initial jet orientation. Thus, while the nature of SN 1998bw remains an open question, there appears to be a clear dichotomy between the majority of hydrodynamic and engine-driven explosions.

SECTION 8.1
Introduction

The death of massive stars and the processes that lead to the formation of the compact remnants is a forefront area in stellar astrophysics. Recent advances in modeling suggest that great diversity can be expected. Indeed, observationally we have already witnessed a large diversity in the neutron star remnants: radio pulsars, AXPs, SGRs, and the central source in Cas A. We know relatively little about the formation of black holes, static or rotating.

[†] A version of this chapter was published in *The Astrophysical Journal*, vol. 599, 408–418, (2003).

The compact objects form following the collapse of the progenitor core. The energy of the resulting explosion can be supplemented, or even dominated, by the energy released from the compact object (e.g., a rapidly rotating magnetar or an accreting black hole). Such "engines" can drive strongly collimated outflows (MacFadyen & Woosley 1999), but even in their absence the core collapse process appears to be mildly asymmetric (e.g., Wang et al. 2001). Regardless of the source of energy, a fraction of the total energy, E_K, is coupled to the debris or ejecta (mass M_{ej}) and it is these two gross parameters which determine the appearance and evolution of the resulting explosion. Equivalently one may consider E_K and the mean initial speed of ejecta, v_0, or the Lorentz factor, $\Gamma_0 = [1 - \beta_0^2]^{-1/2}$, where $\beta_0 = v_0/c$.

Supernovae (SNe) and γ-ray bursts (GRBs), are distinguished by their ejecta velocities. In the former $v_0 \sim 10^4$ km s^{-1} as inferred from optical absorption features (e.g., Filippenko 1997), while for the latter $\Gamma_0 \gtrsim 100$, inferred from the non-thermal prompt emission (Goodman 1986; Paczynski 1986), respectively. The large difference in initial velocity arises from significantly different ejecta masses: $M_{ej} \sim few$ M$_\odot$ in SNe compared to $\sim 10^{-5}$ M$_\odot$ in GRBs.

In the conventional interpretation, M_{ej} for SNe is large because E_K is primarily derived from the mildly asymmetrical collapse of the core and the energy thus couples to most the mass left after the formation of the compact object. Mysteriously, E_K clusters around 1 FOE (FOE is 10^{51} erg) in most SNe, a mere 1% of the energy released in the gravitational collapse of the core.

Whereas the initial ejecta speed is solely determined by E_K and M_{ej}, a fraction of the ejecta is accelerated to higher velocities as the blast wave races down the density gradient of the stellar enveloped (e.g., Matzner & McKee 1999). For the wind- or binary-stripped (e.g., Uomoto 1986; Branch et al. 1990; Woosley et al. 1993; Nomoto et al. 1994) helium and carbon progenitors of Type Ib and Ic SNe, both factors (a smaller core mass and a steep density gradient) conspire to produce ejecta at velocities as high as $\Gamma\beta \sim 1$. However, only $\lesssim 10^{-5} E_K$ is carried by these ejecta (Colgate 1968; Woosley & Weaver 1986; Matzner & McKee 1999). In contrast, high velocity ejecta is neither expected nor observed in Type II SNe with their massive stellar envelopes.

GRB models, on the other hand, appeal to a stellar mass black hole remnant, which accretes matter on many dynamical timescales and powers relativistic jets (the so-called collapsar model; Woosley 1993; MacFadyen & Woosley 1999); highly magnetized neutron stars have also been proposed (e.g., Ruderman et al. 2000). Observationally, this model is supported by the association of some GRBs with SN explosions (e.g., Stanek et al. 2003). In addition, the complex temporal profiles and long duration of GRBs are interpreted in terms of an engine that is relatively long lived (i.e., not a singular explosion). The high Lorentz factors, a high degree of collimation with opening angles of a few degrees (Frail et al. 2001), and episodes of energy addition presumably from shells of ejecta with varying Lorentz factors, further distinguish GRBs from Type Ib/c SNe.

We now recognize that engine-driven events – GRBs and the recently discovered X-ray Flashes – in fact have a wide dispersion in their ultra-relativistic output as manifested by their beaming-corrected γ-ray energies (Bloom et al. 2003b) and X-ray luminosities (Berger et al. 2003a). However, these cosmological explosions appear to have a nearly constant total explosive yield when taking into account the energy in mildly relativistic ejecta (Berger et al. 2003c).

The unusual and nearby ($d \sim 40$ Mpc) SN 1998bw shares some of the unique attributes expected of GRBs. This Type Ic SN was found to be coincident in time and position with GRB 980425 (Galama et al. 1998a), an event with a single smooth profile. The inferred isotropic energy in γ-rays of GRB 980425 was only 8×10^{47} erg (Pian et al. 2000), three to six orders of magnitude fainter than typical GRBs. More importantly, SN 1998bw exhibited unusually bright radio emission indicating about 10^{50} erg of mildly relativistic ejecta (Li & Chevalier 1999). Equally significant, the radio emission indicated a clear episode of energy addition[1] (Li & Chevalier 1999). None of these features – γ-rays, significant energy with $\Gamma\beta \gtrsim 2$, and episodes of energy addition – have been seen in any other nearby SN. Thus, the

---

[1] With the assumption that free-free absorption is the dominant absorption process, the increase in flux has also been interpreted as due to variations in the circumstellar density (Weiler et al. 2001). However, the model proposed by these authors requires unrealistic expansion velocities and/or kinetic energies.

empirical data strongly favor an engine origin for SN 1998bw. In the optical, SN 1998bw also appears to be extreme: velocity widths approaching 60,000 km s^{-1} were seen at early time (Iwamoto et al. 1998) and the inferred explosion energy may be above normal values, with estimates ranging from 2 to 50 FOE (Höflich et al. 1999; Nakamura et al. 2001).

The inference of an engine in SN 1998bw raises two scenarios for its origin and relation to GRBs. GRB 980425 may have been a typical burst but viewed well away from the jet axis (hereafter, the off-axis model), thereby resulting in apparently weak γ-ray emission despite the great proximity. Alternatively, SN 1998bw represents a different class of SNe. If so, collapsars can produce very diverse explosions.

A powerful discriminant between these two scenarios is the expected rate of SN 1998bw-like events. In the off-axis model, the fraction of Type Ib/c SNe that are powered by a central engine is linked to the mean beaming factor of GRBs (e.g., Frail et al. 2001; Totani & Panaitescu 2002). Recently, Frail et al. (2001) presented the distribution of jet opening angles for a sample of 15 GRBs, and found that the mean beaming factor is $\langle f_b^{-1} \rangle \sim 500$; here $f_b = [1 - \cos(\theta_j)]$ is the beaming fraction, and θ_j is the collimation angle. With an estimated local GRB rate of ~ 0.5 Gpc^{-3} yr^{-1} (Schmidt 2001) compared to a Type Ib/c SN rate of $\sim 4.8 \times 10^4$ Gpc^{-3} yr^{-1} (Marzke et al. 1998; Cappellaro et al. 1999; Folkes et al. 1999), we expect that $\sim 0.5\%$ of Type Ib/c SNe will be similar[2] to SN 1998bw.

On the other hand, if SN 1998bw is not an off-axis cosmological burst, then the rate of similar events has to be assessed independent of the GRB rate. An upper limit can be obtained by assuming that all Type Ib/c SNe are engine driven highly asymmetric explosions with SN 1998bw having the most favorable geometry. In this context, Norris (2002) has argued that of the 1429 long-duration BATSE bursts, about 90 events possess similar high-energy attributes as that of GRB 980425. This sub-sample may be concentrated along the super-galactic plane. If this sub-sample is accepted as distinct from the cosmological bursts then $\sim 25\%$ of Type Ib/c SNe within 100 Mpc are expected to be events like SN 1998bw.

Here, we report a comprehensive program of radio observations of nearby Type Ib/c SNe. We began this program in 1999 (motivated by SN 1998bw) and observed most reported Type Ib/c SNe with the Very Large Array. Our basic hypothesis is that (mildly) relativistic ejecta are best probed by radio observations, as was demonstrated in the case of SN 1998bw. Furthermore, radio observations of Type Ib/c SNe allow us to directly compare these objects to the radio afterglows of cosmological GRBs. Thus, we can empirically (direct comparison of radio luminosity distributions) and quantitatively (calorimetry via radio observations) investigate the link, or lack thereof between Type Ib/c SNe and cosmological GRBs. As alluded to above, we did not investigate Type II SNe since the extended envelopes and dense circumstellar media of their progenitors are reasonably expected to mask the activity of a putative engine and thus suppress the presence of mildly relativistic ejecta to which we are most sensitive.

The organization of the paper is as follows. In §8.2 we present the details of the observations. The results are summarized in §8.3, where we investigate the broad radio properties (§8.3.1), expansion velocities (§8.3.2), and energies in high velocity ejecta (§8.3.3). We further provide a comparison to the radio afterglows of GRBs in §8.4 and draw conclusions in §8.5.

SECTION 8.2

Observations

Table 8.1 summarizes the Very Large Array (VLA[3]) observations of Type Ib/c SNe starting in late 1999 and up to the end of 2002. We observed a total of 33 SNe out of 51 identified spectroscopically during the same period. The observed targets were determined solely by the availability of observing time and optical selection criteria; we did not employ any additional selection criteria.

---

[2] We note that this fraction may be somewhat higher in jet models in which the energy and/or Lorentz factor decrease away from the jet axis. The exact fraction depends on the details of the energy and Lorentz factor distribution (Rossi et al. 2002; Zhang & Mészáros 2002).

[3] The VLA is operated by the National Radio Astronomy Observatory, a facility of the National Science Foundation operated under cooperative agreement by Associated Universities, Inc.

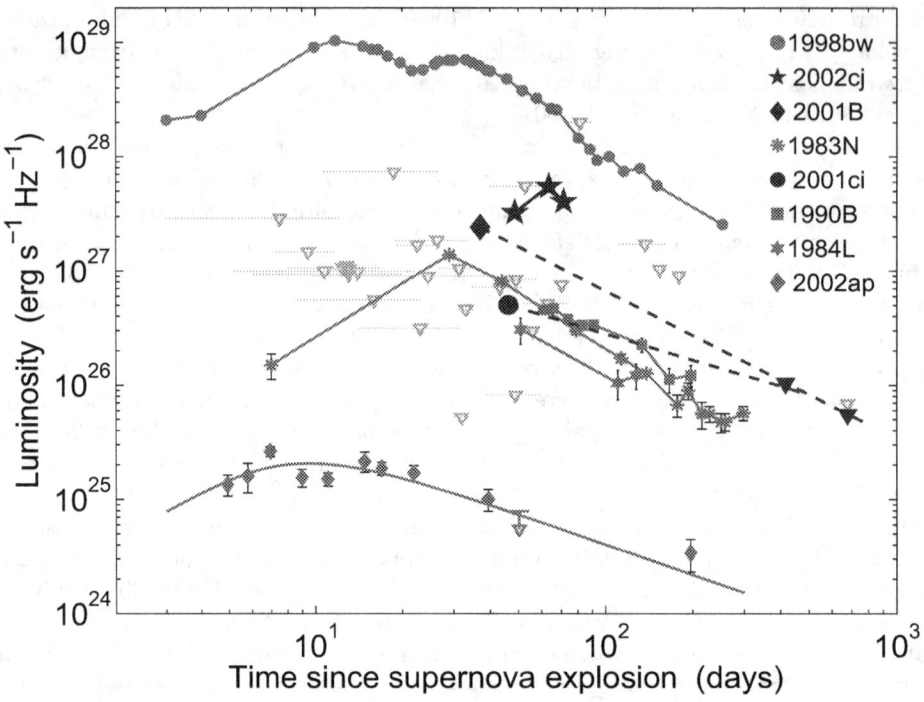

Figure 8.1: Radio light curves of Type Ib/c SNe detected in this survey and from the literature, as well as upper limits for the non-detections (triangles); these are plotted as 3σ in most cases with the exception of SNe which are located on top of a bright host galaxy (see Table 8.1). The light curves are at 8.5 GHz (SN 1998bw), 4.9 GHz (SNe 1983N, 1984L, and 1990B), and 1.4 GHz (SN 2002ap); these frequencies were chosen since they provide the best coverage of the radio evolution. For SN 2002ap we plot the model of Berger et al. (2002b), while the other solid lines simply trace the observations and do not represent a model fit. The uncertainty in time for the non-detections represents the uncertain time of explosion. We note that for SN 2002cg, which is the only SN that is potentially brighter than SN 1998bw, the limit is 10σ due to the superposition of the SN on top of its host galaxy.

In all observations we used the standard continuum mode with 2×50 MHz bands, centered on 1.43, 4.86, or 8.46 GHz. We used the sources 3C 48 (J0137+331), 3C 147 (J0542+498), and 3C 286 (J1331+305) for flux calibration, and calibrator sources within $\sim 5°$ of the SNe to monitor the phase. The data were reduced and analyzed using the Astronomical Image Processing System (Fomalont 1981).

┌─ SECTION 8.3 ───┐

Population Statistics

└───┘

In this section we investigate the ejecta properties and diversity of the sample. Results for individual SNe are given in the Appendix. In Figure 8.1 we plot the radio luminosities and upper limits for Type Ib/c SNe observed in this survey and in the past (SN 1983N: Sramek et al. 1984; SN 1984L: Panagia et al. 1986; Weiler et al. 1986; SN 1990B: van Dyk et al. 1993; SN 1998bw: Kulkarni et al. 1998; SN 2002ap: Berger et al. 2002b). The typical delay between the SN explosion and time of our observations is about 20 days, with four SNe observed with a delay of over 100 days. In addition, three of the SNe are embedded in host galaxies with strong radio emission. For these cases, we adopt upper limits that correspond to the brightness of the galaxy (at least ten times the root-mean-square noise of the individual image). Four of the thirty three SNe have been detected. Thus the detection rate of our

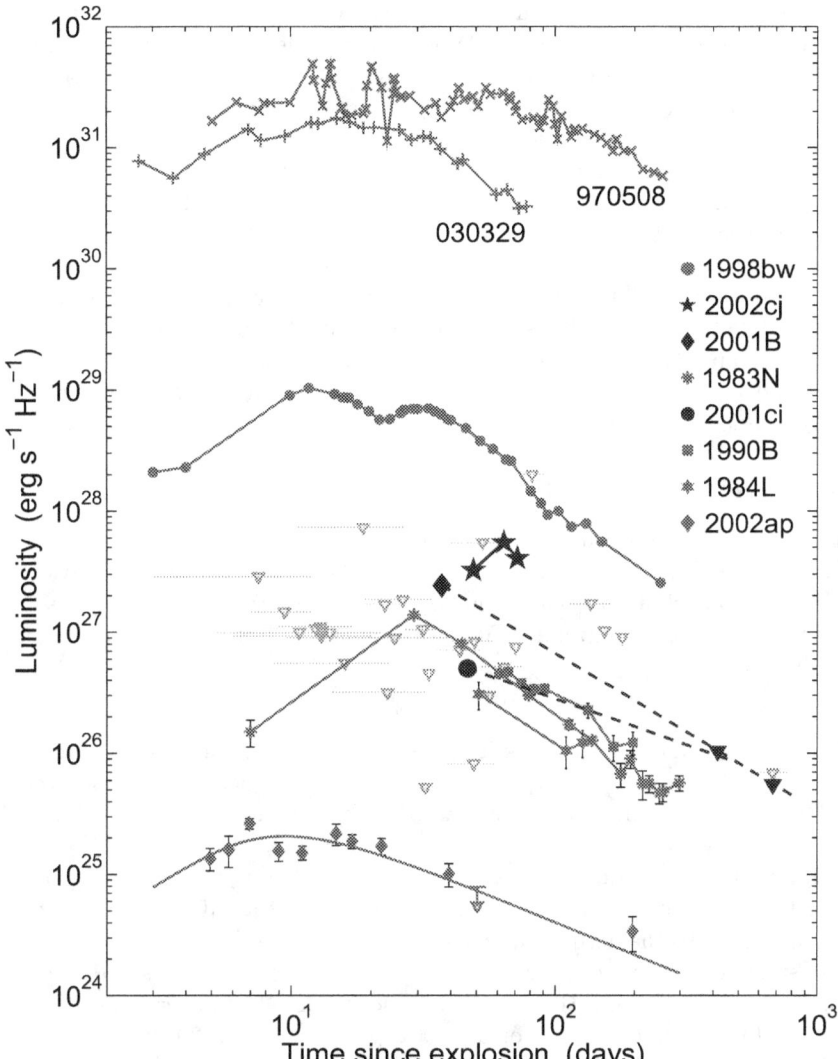

Figure 8.2: Same as Figure 8.1 but including the 8.46 GHz light curves of GRB 970508 (Frail et al. 2000c) and GRB 030329 (Berger et al. 2003c). These GRB afterglows are at least two orders of magnitude brighter than SN 1998bw, the brightest Type Ib/c SN. We note that the fluctuations in the GRB light curves are not intrinsic, and arise instead from interstellar scintillation. Based on the significant difference in radio luminosity we rule out the possibility that the Type Ib/c SN observed here produced a GRB. This is discussed more quantitatively in in §8.4 and Figures 8.4 and 8.5.

experiment with a typical flux density limit of 0.15 mJy (3σ) is about 12%.

8.3.1 Radio Properties of Type Ib/c SNe

Figure 8.1 provides a succinct summary of the radio light curves of the Type Ib/c SNe. Two strong conclusions can be immediately drawn from this Figure. First, SNe as bright as SN 1998bw are rare. Second, there is significant dispersion in the luminosities of Type Ib/c SNe, ranging from $L_{\nu,\mathrm{rad}} \approx 10^{29}$ erg s^{-1} Hz^{-1} at the bright end (SN 1998bw) to that of SN 2002ap which is fainter by about four orders of magnitude (Berger et al. 2002b). It is curious that SN 2002ap also happens to be the nearest Ib/c SN in our sample (Table 8.1). Six of the eight Type Ib/c SNe detected in the radio to date cluster in the range of about $(3 - 50) \times 10^{26}$ erg s^{-1} Hz^{-1}. This may be partly due to a selection effect since the typical detection threshold is about $4 \times 10^{26}(d/50\,\mathrm{Mpc})^2$ erg s^{-1} Hz^{-1}.

We also find that 28 of the 29 non-detections are no brighter than 0.1 times the luminosity of SN 1998bw. SN 2002cg appears potentially brighter than SN 1998bw only because it is embedded in a radio bright host galaxy; we are therefore forced to use a 10σ limit on its luminosity (Table 8.1). Thus, the incidence of bright events like SN 1998bw is $\lesssim 3\%$.

As with the radio luminosities, the peak times also exhibit great variation: at 1.4 GHz the emission from SN 2002ap peaked at about 7 days, while for SN 2002cj it peaked at about 65 days. For SN 1998bw the initial peak occurred at 15 days, followed by a second peak at about 40 days. Similarly, at 8.5 GHz, SN 1998bw peaked at 12 and 30 days past explosion, SN 1983N peaked at about 30 days, and SN 2002ap is predicted to have peaked at $\sim 1 - 2$ days (the first observation at this frequency was taken about 4 days after the SN explosion).

8.3.2   Expansion Velocities

If the radio emission arises from a synchrotron spectrum peaking at the self-absorption frequency, ν_a, then the peak time and peak luminosity directly measure the mean expansion speed (Chevalier 1998). This is simply because the self-absorption frequency is sensitive to the size of the source, while the luminosity is sensitive to the swept-up mass. We use Equation 16 of Chevalier (1998) to evaluate the average expansion velocities:

$$v_p \approx 3.1 \times 10^4 L_{p,26}^{17/36} t_{p,10}^{-1} \nu_{p,5}^{-1} \text{ km s}^{-1}. \tag{8.1}$$

Here, $L_p = 10^{26} L_{p,26}$ erg s^{-1} Hz^{-1} is the peak luminosity, $t_p = 10 t_{p,10}$ days is the time of peak emission relative to the SN explosion, and $\nu_p = 5\nu_{p,5}$ GHz is the peak frequency. We infer velocities ranging from $v \sim 10^4$ to 10^5 km s^{-1} (Figure 8.3). Again, as with the luminosities, SN 1998bw with $v \sim c$ is an exception.

We note that if free-free absorption plays a significant role, then ν_p is only an upper limit to ν_a, and L_p is a lower limit to the intrinsic peak luminosity. In this case, the inferred values of v_p listed above will in fact be a lower limit to the actual expansion velocity. However, this is probably not significant for Type Ib/c SNe since their compact progenitors have high escape velocities and therefore fast winds and low circumburst densities. Indeed, there is no evidence for free-free absorption either for SN 1998bw (Kulkarni et al. 1998; Li & Chevalier 1999) or SN 2002ap (Berger et al. 2002b).

Estimating the expansion velocities for the non-detections is not straightforward since we cannot ensure that the limits constrain the peak luminosity. We are therefore forced to make an additional assumption. For example, if we assume that most Type Ib/c SNe are similar in their emission properties to SN 1983N, then the majority of upper limits approximately sample the peak emission and the inferred upper limits are $\lesssim 0.3c$ (Figure 8.3).

On the other hand, if the typical peak time is only a few days then our observations only constrain the decaying portion of the light curve and the expansion velocity may be higher. Fortunately, this is not a significant problem based on the following argument. The equipartition energy directly depends on the peak luminosity, $U_{\mathrm{eq}} \approx 3.7 \times 10^{46} L_{p,27}^{20/17}$ erg (see §8.6.3), where $L_p = 10^{27} L_{p,27}$ erg s^{-1} is the peak luminosity at 8.5 GHz. With a typical fading rate of $F_\nu \propto t^{-1}$ in the optically thin regime, there are a few SNe that could have reached a peak luminosity of the order of 10^{29} erg s^{-1} if $t_p = 1$ day post explosion. This is a reasonable minimum peak time taking into account the deceleration time of the ejecta. Thus, the equipartition energy is at most 10^{49} erg, about two orders of magnitude lower than typical GRBs (§8.5). For most non-detections the limit is in fact much lower, $\sim 5 \times 10^{46}$ to 10^{48} erg. This indicates that a few of the non-detected sources may have in fact produced mildly relativistic ejecta, but these would still be energetically uninteresting when compared to SN 1998bw let alone GRB afterglows.

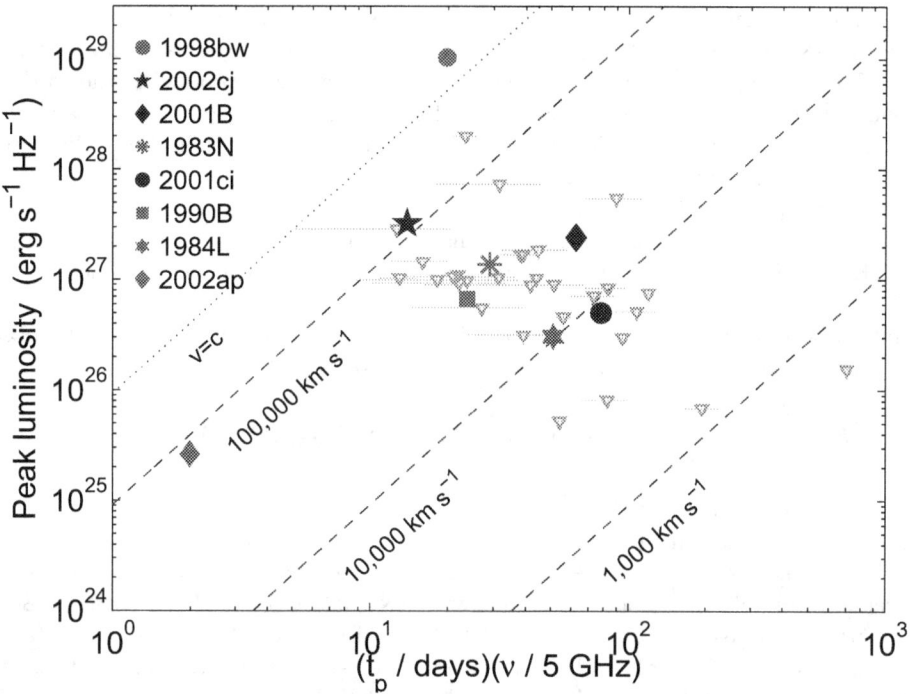

Figure 8.3: Peak radio luminosity plotted versus the time of peak luminosity for Ib/c SNe studied in this survey and from the literature. Symbols are as in Figure 8.1. The diagonal lines are contours of constant average expansion velocity based on the assumption that the peak of the radio luminosity occurs at the synchrotron self-absorption frequency (Chevalier 1998). While the upper limits do not necessarily measure the peak of the spectrum at the time of the observation, a comparison to SN 1983N indicates that the range of time delays relative to the SN explosion reasonably samples the peak. Upper limits measured at $t \gtrsim 100$ days probably miss the peak of the synchrotron spectrum and therefore do not provide a useful limit.

8.3.3 Energetics

In the previous section we found that no SN observed to date is comparable to SN 1998bw especially in regard to the mean expansion speed. SN 1998bw was also interesting because it possessed an unusually large amount of energy in mildly relativistic ejecta. However, a purely hydrodynamic explosion can also produce *some* amount of relativistic ejecta. The energy of such ejecta can be estimated using well understood models of shock propagation in the pre-supernova cores (Chevalier 1982; Matzner & McKee 1999). The key parameters are E_K and M_{ej} which can be inferred from the optical light curves and spectra using hydrodynamic models of a SN explosion in a CO core coupled with radiative transfer calculations (e.g., Iwamoto et al. 2000).

In this section, we investigate whether any of the detected Type Ib/c SNe possess such large energy in high velocity ejecta that cannot be explained by the simplest hypothesis of a purely hydrodynamic explosion. To this end, in Table 8.2 we summarize the results of hydrodynamic models for the SNe that have been detected in the radio.

The ejecta produced in a hydrodynamic explosion has a density profile that can be described by power laws at low and high velocities, separated by a break velocity, which for Type Ib/c progenitors is given by (Matzner & McKee 1999):

$$v_{ej,b} \approx 5.1 \times 10^3 (E_{K,51}/M_{ej,1})^{1/2} \, \text{km s}^{-1}. \tag{8.2}$$

Here $E_K = 10^{51} E_{K,51}$ erg and $M_{\rm ej} = 10 M_{\rm ej,1}$ M$_\odot$. For typical values of E_K and $M_{\rm ej}$, the radio emission from the detected SNe is produced by ejecta above the break velocity. In particular, for SN 2002ap, $v_{\rm ej,b} \approx 2 \times 10^4$ km s^{-1}, which is lower than the velocity of the ejecta producing the radio emission, $v \approx 9 \times 10^4$ km s^{-1} (Berger et al. 2002b). Similarly, for SN 1998bw $v_{\rm ej,b}$ ranges from about 1.5×10^4 to 3.5×10^4 km s^{-1} (depending on which model is assumed, Table 8.2) while the radio emission was produced by ejecta expanding with $\Gamma\beta \approx 2$.

The ejecta velocity profile extends up to a cutoff determined by significant radiative losses when the shock front breaks out of the star. For a radiative stellar envelope this is $v_{\rm ej,max} \approx 11.5 \times 10^4 E_{K,51}^{0.58} M_{\rm ej,1}^{-0.42}$ (Matzner & McKee 1999), assuming a stellar radius of 1 R$_\odot$. For the SNe considered here we find cutoff velocities of $\Gamma\beta \sim 1 - 3$.

To determine whether there is sufficient energy in fast ejecta to account for the radio observations we calculate the energy above a velocity, V (Matzner & McKee 1999):

$$E(v > V) \approx \int_V^\infty \frac{1}{2}\rho_f v^2 4\pi v^2 t^3 dv \approx 7.2 \times 10^{44} E_{K,51}^{3.59} M_{\rm ej,1}^{-2.59} V_5^{-5.18} \text{ erg}, \qquad (8.3)$$

where V_5 is the velocity in units of 10^5 km s^{-1}.

Unfortunately, as can be seen from Table 8.2, only four (including SN 1998bw) SNe have sufficient optical data which is necessary to estimate E_K and $M_{\rm ej}$. Of this limited sample, much of the radio data for SN 1994I remain unpublished. Thus we are left with SN 2002ap, SN 1983N, and SN 1998bw.

Using the parameters given in Table 8.2 for SN 2002ap, Berger et al. (2002b) find $E(v > 0.3c) \approx 3.8 \times 10^{48}$ erg. In contrast, from the radio observations we estimate 2×10^{46} erg. Thus, there is no need, nor indeed room, for mildly relativistic ejecta in this SN. We therefore disagree with the claims of high velocity jets carrying a large amount of energy, $\sim 0.1 - 1$ FOE, made by Kawabata et al. (2002) and Totani (2003). Furthermore, the large discrepancy between the amount of energy inferred from the hydrodynamic models and the radio observations suggests that either the optically-derived parameters are in error, Equation 8.3 has an incorrect pre-factor, or the radio estimate is incorrect. However, the radio estimate is relatively robust (eventually related to equipartition energy estimates). On the other hand, as with SN 1998bw (Höflich et al. 1999) the total kinetic energy may have been over-estimated, possibly as a result of neglecting a mild asymmetry.

For SN 1984L we do not have a direct estimate of the energy in the radio-emitting ejecta since the peak of the radio emission has been missed. However, based on the similarity to SN 1983N in the optically thin regime we estimate $L_p(t = 30 \, {\rm d}, \nu_p = 5 \, {\rm GHz}) \approx 1.4 \times 10^{27}$ erg s^{-1} Hz^{-1}. This translates to a peak flux of 3.2 mJy at the distance of SN 1984L ($d \approx 19$ Mpc). Using the equipartition analysis presented in §8.6.3 we estimate an energy of about 7×10^{46} erg, and an average expansion velocity of about $0.1c$. From Equation 8.3 we find $E(v > 0.1c) \approx 3 \times 10^{50}$ erg – similar to the conundrum discussed above for SN 2002ap.

For SN 1998bw, on the other hand, we find $E(v > c) \approx 2 \times 10^{45}$ erg using the parameters inferred by Höflich et al. (1999), or $E(v > c) \approx 3 \times 10^{48}$ erg using the parameters given by Iwamoto et al. (1998). In both cases, the energy available in fast ejecta is significantly lower than the energy inferred from the radio emission, $\sim 10^{50}$ erg.

To conclude, for SN 1984L and SN 2002ap a hydrodynamic explosion can supply the energy and velocity that are responsible for the observed radio emission. Most likely, the same is true for the non-detections. On the other hand, SN 1998bw is a clear exception, exhibiting a significant excess of energy in ejecta moving with $\Gamma\beta \approx 2$ compared to what is available from a hydrodynamic explosion.

SECTION 8.4

A Comparison to γ-Ray Burst Afterglows

In the previous section we investigated the radio properties of Type Ib/c SNe and found that in every respect SN 1998bw was unique. In this section we compare the Ib/c sample (including SN 1998bw)

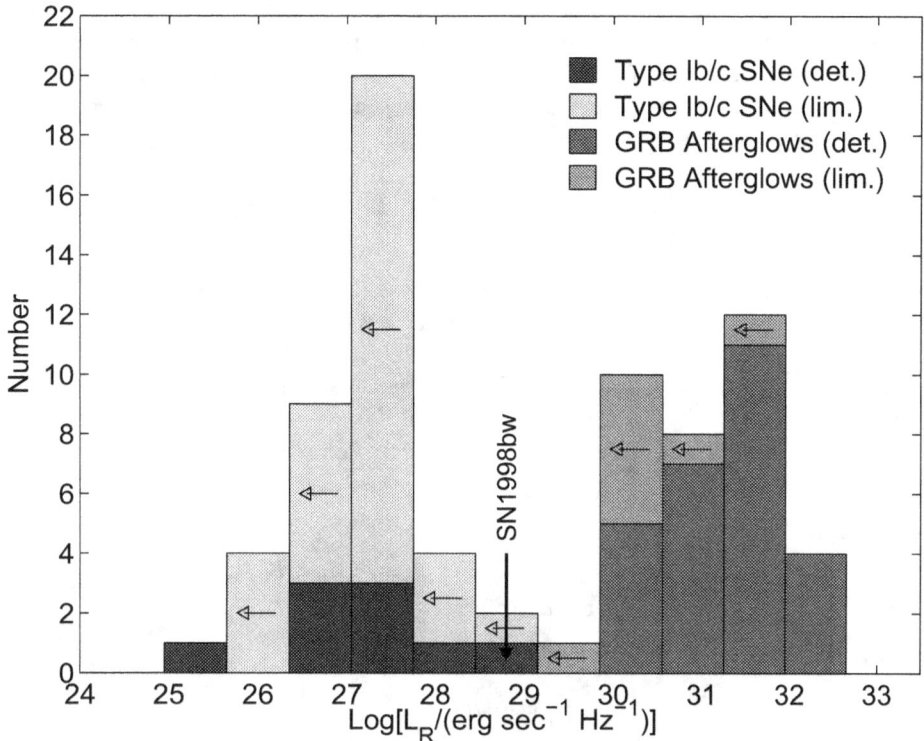

Figure 8.4: Histograms of the radio luminosity of Type Ib/c SNe from this survey and the literature, and GRB radio afterglows from the sample of Frail et al. (2003). Upper limits are plotted as 3σ, unless there is significant contamination from the host galaxy (see Table 8.1).

with the radio afterglows of GRBs. In Figure 8.2 we plot the radio light curves of GRB 970508 (Frail et al. 2000c), a typical cosmological burst, and the nearest event, GRB 030329 (Berger et al. 2003c), in addition to the SN light curves. As demonstrated by this figure and Figure 8.4, the radio light curves of GRB afterglows and SNe are dramatically different. Furthermore, SN 1998bw is unique in both samples: it is fainter than typical radio afterglows of GRBs but much brighter than Type Ib/c SNe.

Figures 8.2 and 8.4 have significant implications, namely, *none* of the Type Ib/c SNe presented here could have given rise to a typical γ-ray burst. It has been suggested that GRBs are distant Type Ib/c SNe but with their jets pointed at the observer, whereas such a bias is absent in the nearby Type Ib/c sample. However, most of our radio observations are obtained on a timescale of 10–100 days (see Figure 8.1). Scaling from the observed "jet" break times of a few days in GRB afterglows, off-axis collimated explosions become spherical on a timescale of $\sim 10 - 10^2$ days (Paczynski 2001; Granot & Loeb 2003) at which point the relative geometry between the observer and the explosion is not important. Thus, we find no evidence suggesting that all or even a reasonable majority of Type Ib/c SNe give rise to GRBs. We now quantify the difference between the Ib/c and GRB samples.

Our goal here is to determine the differential luminosity distribution, $n(L)$, which agrees with both detections and upper limits; here $n(L)$ is the number of events with luminosity between L and $L + dL$. It is important to include non-detections since they represent the majority of the SN data. Similarly, we include upper limits on the radio luminosity of GRB afterglows that have been localized in other wave-bands (i.e., optical and X-rays) and for which a redshift has been measured. Unfortunately, as many as half of the GRBs localized in the X-rays do not have a precise position, and hence a redshift. For these afterglows it is not possible to provide a limit on the radio luminosity. Still, with a typical flux limit of about 0.3 mJy (5σ; Frail et al. 2003a), and assuming that these sources have a similar distribution of redshifts to the detected afterglows, we find typical luminosity limits of about 10^{31} erg s^{-1} Hz^{-1}, consistent with the peak of the distribution of detected afterglows. Therefore, unless these sources are biased to low redshift we do not expect a strong bias as a result of neglecting them.

The quality of fit for $n(L)$ is determined using the Likelihood function, $\mathcal{L} = \prod_{i=1}^{N} \mathcal{L}_i$, with (Reichart & Yost 2001):

$$\mathcal{L}_i = \begin{cases} \int_{-\infty}^{\infty} n(L)\mathcal{G}(L_i, \sigma_{Li})dL & L_i = detection \\ \int_{L_i}^{\infty} n(L)dL & L_i = limit, \end{cases} \tag{8.4}$$

where N is the total number of sources (SNe or afterglows), and $\mathcal{G}(L_i, \sigma_{Li})$ is a normalized Gaussian profile centered on the observed luminosity of a detected source and with a width equal to the 1σ rms uncertainty in the luminosity.

We consider four models for $n(L)$ based on the apparent distribution of the detections and upper limits: a Gaussian,

$$n(L) = \frac{1}{\sqrt{2\pi}\sigma_L} \exp\left[-\frac{1}{2}\left(\frac{L - L_0}{\sigma_L}\right)^2\right], \tag{8.5}$$

a decreasing power-law,

$$n(L) = \begin{cases} 0 & L < L_0 \\ (1 - \alpha_L)L^{\alpha_L}/(L_0^{\alpha_L + 1}) & L \geq L_0, \end{cases} \tag{8.6}$$

an increasing power-law,

$$n(L) = \begin{cases} (1 + \alpha_L)L^{\alpha_L}/(L_0^{\alpha_L + 1}) & 0 \leq L < L_0 \\ 0 & L \geq L_0, \end{cases} \tag{8.7}$$

and a flat distribution,

$$n(L) = \begin{cases} 0 & L < L_1 \\ 1/(L_2 - L_1) & L_1 \leq L \leq L_2 \\ 0 & L > L_2. \end{cases} \tag{8.8}$$

In each case we fit for the two free parameters (e.g., L_0 and σ_L in Equation 8.5). We do not use the increasing power law model for the individual distributions since the observations are clearly inconsistent with such a model. The resulting best-fit models are shown in Figure 8.5 and summarized in Table 8.3.

We find that the SN population is modeled equally well with the Gaussian, flat, or decreasing power law distributions, while the GRB afterglows can be fit with a Gaussian or flat distributions; a decreasing power law provides a much poorer fit. Regardless of the exact distribution the two populations require distinctly different parameters, with a minimal overlap at the tails of the distributions.

Fitting the combined SN and afterglow data with the models provided above (Figure 8.5 and Table 8.3) we find that even the best models (an increasing power-law or a flat distribution) provide a much poorer fit; the likelihood of the fits are $\ln(\mathcal{L}) \approx 104$ compared to the combined value of 61.3 for the separate Gaussian fits. Thus, the two populations cannot be accommodated with a simple single distribution. This points to a separate origin for the GRB and Type Ib/c SN populations. However, SN 1998bw can be accommodated in either population. It is equally plausible that it is a low luminosity GRB or the brightest Type Ib/c radio supernova known to date.

SECTION 8.5

Discussion and Conclusions

We presented VLA radio observations of 33 Type Ib/c SNe observed between late 1999 and the end of 2002. Four of these SNe have been detected, giving a detection rate of about 12% above a typical 3σ

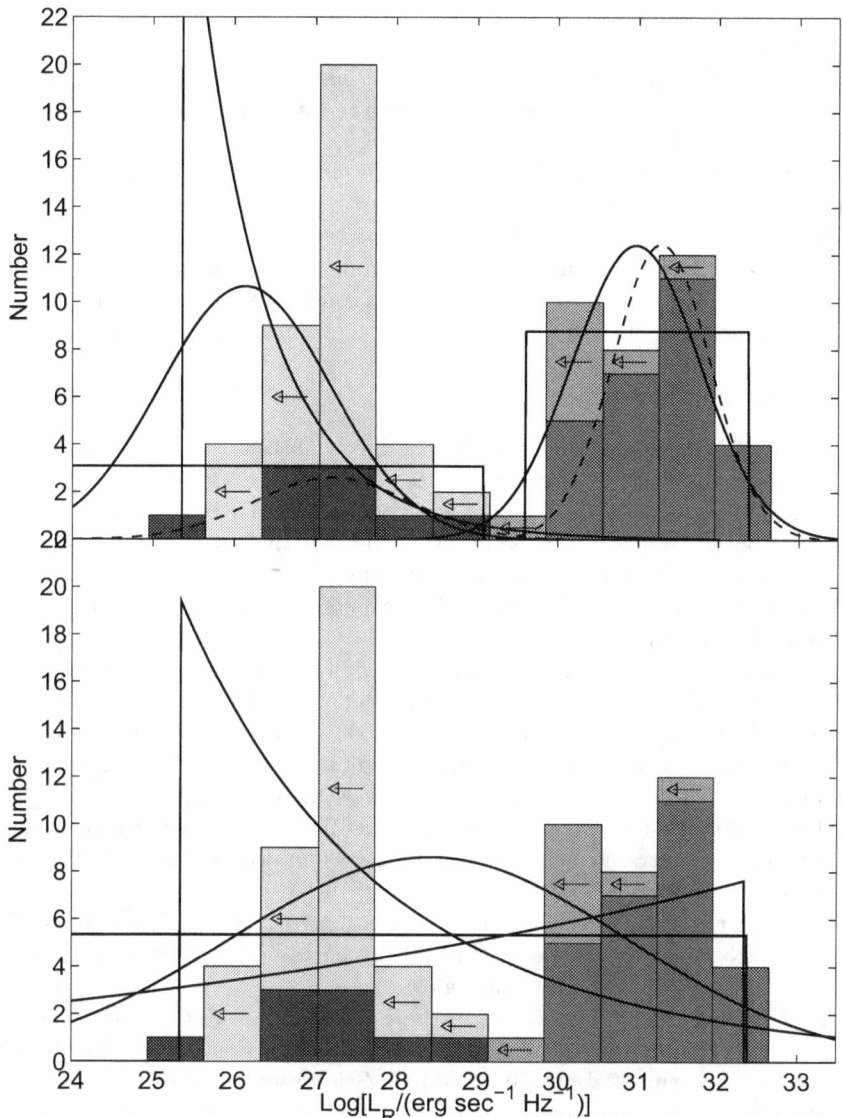

Figure 8.5: Same as Figure 8.4 but with models of the luminosity distribution (§8.4). The dashed lines are a Gaussian profile fit for the detections in each sample. The solid lines are a fit to the detections and upper limits using several models for the distribution function (§8.4). In the top panel we model each population separately, whereas the bottom panel shows models for the combined populations. No single distribution can fit both the local Type Ib/c SN population and the cosmological GRB population (Table 8.3).

flux limit of 0.15 mJy. At the same time, the combined detections and non-detections indicate that at most 3% of Type Ib/c SNe are as luminous as SN 1998bw, although the single source which may be brighter is only so because it is embedded in a radio bright host galaxy.

We infer typical velocities of the radio-emitting ejecta of about $10^4 - 10^5$ km s^{-1} for the detected SNe, and upper limits in the same range for the non-detections. We also find that a hydrodynamic explosion can supply the energy carried by the fastest ejecta. Finally, none of the detected SNe show clear evidence for variable energy input (shells with different velocity or continued activity by the central engine); however, we note that our sampling is quite sparse.

The measurements (radio light curves) and inferences (energy in fast ejecta, energy addition) offer no compelling reason to conclude that any of our SNe have the special properties of SN 1998bw (§8.5.1).

Norris (2002) has proposed, based on the empirical lag-luminosity relation, that 25% of Type Ib/c SNe are similar to SN 1998bw. Clearly, this conclusion is not borne out by observations. This may indicate that the lag-luminosity relation does not apply to bursts with long lags (i.e., low luminosity) which comprise the local sample of Norris (2002)

We also compared the Type Ib/c sample with the sample of radio afterglow of GRBs. Empirically, these two populations appear to be quite disparate. This conclusion is reinforced by careful modeling of the luminosity distributions. Still, SN 1998bw may belong to either population.

8.5.1   What Is SN 1998bw?

Our four year survey of Type Ib/c SNe was first and foremost motivated by the peculiar object, SN 1998bw. This supernova showed three attributes unique to GRBs: relativistic ejecta, substantial reservoir of energy in such ejecta, and energy addition. A singular hydrodynamic explosion cannot account for these attributes. A natural explanation is that SN 1998bw, like GRBs, was driven by an engine powerful enough to significantly modify the explosion.

Our survey has demonstrated that SN 1998bw-like events are rare in the local sample. This begs the question: what is SN 1998bw?

Two popular scenarios have been suggested. The first – the "off-axis" scenario – holds that SN 1998bw is a typical GRB albeit nearby and with collimated ejecta pointed away from us (MacFadyen & Woosley 1999; Nakamura 1999; Granot et al. 2002). This hypothesis is attractive because of its simplicity. We know GRBs exist and most of them do not point towards us (Frail et al. 2001).

In the other scenario SN 1998bw is a new type of explosion (GRB supernovae, or gSN) with little energy in ultra-relativistic ejecta (Bloom et al. 1998b; Kulkarni et al. 1998; Höflich et al. 1999). Evidence in favor of this idea is best illustrated by Figure 8.6 where we find that GRB 980425 is consistently at the faint end of the GRB population.

Unfortunately, we are not able to decisively resolve this controversy. As demonstrated by Figure 8.4, one could argue that SN 1998bw is at the bright end of the radio luminosity function of Type Ib/c supernovae or at the faint end of GRB radio afterglow.

The expected rate of SN 1998bw-like events in the off-axis framework is about 0.5% of Type Ib/c events, given the average beaming factor of about 500 derived by Frail et al. (2001). Thus, it is not entirely improbable that one out of about 40 Type Ib/c SNe observed to date is an off-axis GRB. As an aside, we can use our 3% limit and compare the event rate of Type Ib/c SNe with the observed rate of GRBs (§8.1) to place a limit of $f_b \gtrsim 3 \times 10^{-4}$ on the beaming fraction. This corresponds to a limit of $\theta_j \gtrsim 1.4°$ on the jet opening angles of GRBs; narrower jets are not likely. This result may also be interpreted as a limit on angular size of the highly relativistic core in models of variable energy and/or Lorentz factor across the surface of the jet (Rossi et al. 2002; Zhang & Mészáros 2002).

In the gSN framework, we now know that at most a few percent of Type Ib/c are possibly gSNe. At the same time, the recent GRB 030329 was accompanied by a SN similar to SN 1998bw (SN 2003dh; Stanek et al. 2003; Hjorth et al. 2003). Thus, an investigation of the number and properties of gSNe requires observations of both local Type Ib/c SNe and GRBs.

While we cannot determine the exact origin of SN 1998bw based on the statistics of our survey, the ultimate detection of similar events at the level of about 1% may in fact allow us to distinguish between the off-axis and gSN scenarios.

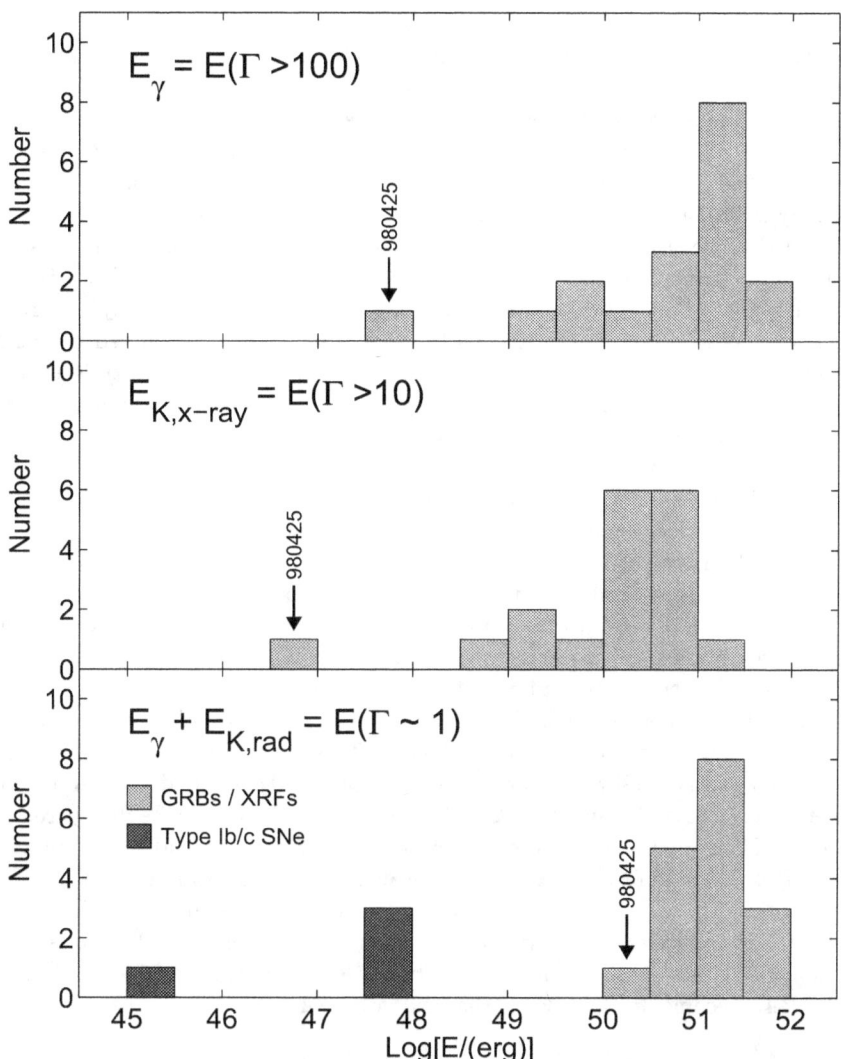

Figure 8.6: Histograms of the beaming-corrected γ-ray energy (Bloom et al. 2003b), E_γ, the kinetic energy inferred from X-rays at $t = 10$ hr (Berger et al. 2003a), $E_{K,X}$, and total relativistic energy, $E_\gamma + E_K$, where E_K is the beaming-corrected kinetic energy inferred from the broad-band afterglows of GRBs (Li & Chevalier 1999; Panaitescu & Kumar 2002) and radio observations of SNe. The wider dispersion in E_γ and $E_{K,X}$ compared to the total energy indicates that engines in cosmic explosions produce approximately the same quantity of energy, thus pointing to a common origin, but the ultra-relativistic output of these engines varies widely. In Type Ib/c SNe, on the other hand, the total explosive yield in fast ejecta (typically $\sim 0.3c$) is significantly lower. This points to a separate origin for these two explosive phenomena.

8.5.2   Hypernovae

The discovery of broad optical lines in SN 1998bw and large explosive energy release, $\gtrsim$ few FOE, prompted some astronomers to use the designation "hypernovae" for SN 1998bw-like SNe. Unfortunately, this designation is not well defined. To begin with, the term hypernova was first used by Paczynski (1998) to describe the GRB/afterglow phenomenon; thus, this term implies a connection to

GRBs. The prevalent view now is that hypernovae are characterized by broad optical absorption lines and larger than normal energy release. However, neither of these criteria has been defined quantitatively by their proponents.

Ignoring this important issue, the following have been suggested to be hypernovae: the Type Ib/c SNe 1992ar (Clocchiatti et al. 2000), 1997dq (Matheson et al. 2001a), 1997ef (Iwamoto et al. 2000; Mazzali et al. 2000), 1998ey (Garnavich et al. 1998), and 2002ap (Mazzali et al. 2002), and the Type II SNe 1992am (Hamuy 2003), 1997cy (e.g., Germany et al. 2000), and 1999E (Rigon et al. 2003). Some have also been claimed to be associated with GRBs detected by BATSE, but at a low significance.

Our view is that the critical distinction between an ordinary supernova and a GRB explosion are relativistic ejecta carrying a considerable amount of energy. Such ejecta are simply not traced by optical spectroscopy. This reasoning is best supported by the fact that the energy carried by the fast ejecta in SN 1998bw and SN 2002ap (Berger et al. 2002b) differ by four orders of magnitude even though both exhibit broad spectral features at early times (see also Wang et al. 2003). Thus, broad lines do not appear to be a good surrogate for SN 1998bw-like objects.

In addition, in two cases, SNe 2002ap and 1984L, the energy inferred from the radio observations indicates that the total kinetic energy as inferred by optical spectroscopy and light curves may have been over-estimated by an order of magnitude (§8.3.3). It is possible that "hypernovae" are in fact only slightly more energetic than typical Type Ib/c SNe, but exhibit a mild degree of asymmetry, leading to excessively high estimates of the total energy when a spherical explosion is assumed.

We suggest that the term hypernova be reserved for those SNe, like SN 1998bw, which show direct evidence for an engine through the presence of relativistic ejecta. As illustrated by SN 1998bw, the relativistic ejecta are reliably traced by radio observations.

We end with the following conclusions. First, radio observations provide a robust way of measuring the quantity of energy associated with high velocity ejecta. This allows us to clearly discriminate between engine-driven SNe such as SN 1998bw and ordinary SNe, powered by a hydrodynamic explosion, such as SN 2002ap (Berger et al. 2002b) and SNe 2001B, 2001ci, and 2002cj presented here. Second, since at least 97% of local Type Ib/c SNe are not powered by inner engines and furthermore have a total explosive yield of only 10^{48} erg in fast ejecta (where fast means $v \sim 0.3c$ compared to $\Gamma \sim few$ in GRBs), there is a clear dichotomy between Type Ib/c SNe and cosmic, engine-driven explosions (Figure 8.6). The existence of intermediate classes of explosions and the nature of SN 1998bw may be ascertained with continued monitoring of several hundred Type Ib/c SNe and cosmological explosions. Fortunately, such samples will likely become available over the next few years.

┌─ SECTION 8.6 ──

Results for Individual Supernovae

└──

8.6.1 SN 2001B

SN 2001B was discovered in images taken on 2001, Jan 3.61 and 4.57 UT, approximately 5.6 arcsec west and 8.9 arcsec south of the nucleus of IC 391 (Xu & Qiu 2001). The SN explosion occurred between 2000, Dec 24.54 UT and the epoch of discovery. Based on an initial spectrum, taken on 2001, Jan 14.18 UT Matheson et al. (2001b) concluded that the SN was of Type Ia approximately 9 days past maximum brightness, showing well defined Si II and Ca II features, with a Si II expansion velocity of about 7400 km sec^{-1}. A subsequent spectrum taken on 2001, Jan 23 indicated that the SN was in fact of Type Ib based on clear He I absorption lines (Chornock & Filippenko 2001).

We observed SN 2001B on 2001, Feb 4.49 UT at 8.46 GHz. An initial analysis revealed a source which was interpreted as the host galaxy of the SN, with a flux of 3.5 mJy. A second epoch obtained on 2002, Oct 28.45 UT revealed that the source had faded below 0.12 mJy, establishing it as the radio counterpart of SN 2001B. (Figure 8.1).

8.6.2   SN 2001ci

SN 2001ci was initially detected in images taken with the Katzman Automatic Imaging Telescope on 2001 Apr 25.2 UT, 6.3 arcsec west and 25.4 arcsec north of NGC 3079 (Swift et al. 2001). These observations did not provide conclusive evidence that the source was in fact a SN. A spectrum obtained by Filippenko & Chornock (2001) on 2001, May 30 UT revealed that the source was in fact a Type Ic SN about 2 − 3 weeks past maximum brightness (Matheson et al. 2001a).

We observed the SN on 2001, Jun 6.35 UT at 8.46 GHz, but did not detect the source since it appeared to be part of the host galaxy extended structure. A second epoch taken on 2002, Jun 10.9 UT revealed a clear fading at the optical position of the SN with a flux of 1.45 ± 0.25 mJy in the first epoch, and a 2σ limit of 0.3 mJy in the second epoch.

8.6.3   SN 2002cj

SN 2002cj was discovered on 2002, Apr 21.5 UT, 1.4 arcsec west and 3.9 arcsec south of the nucleus of ESO 582-G5, at a distance of 106 Mpc (Ganeshalingam & Li 2002). The SN explosion occurred between Apr 9.5 UT and the epoch of discovery. Spectra of the SN obtained on 2002, May 2.43 and May 7 UT revealed that SN 2002cj is of Type Ic (Matheson et al. 2002; Chornock 2002).

We initially observed the SN on 2002, Jun 3.19 UT at 1.43, 4.86, and 8.46 GHz and detected a faint source at all three frequencies. The SN spectrum is given by $F_\nu \propto \nu^{-0.1\pm0.3}$ between 1.43 and 4.86 GHz, and $F_\nu \propto \nu^{-1.2\pm0.8}$ between 4.86 and 8.46 GHz, indicating that the spectral peak was most likely located between 1.43 and 4.86 GHz during the first epoch ($\Delta t \approx 43 - 55$ days). In subsequent observations at 1.43 GHz the SN brightened and then faded, as expected if the peak was in fact above 1.43 GHz initially, and shifted through the band at later epochs. Using the expected shape of the spectrum, $F_\nu \propto \nu^{5/2}$ for $\nu < \nu_p$ and $F_\nu \propto \nu^{-(p-1)/2}$ for $\nu > \nu_p$ (with $p \sim 3$), we find $F_{\nu,p} \sim 0.5$ mJy at $\nu_p \sim 2$ GHz and $\Delta t = 43 - 55$ days.

We use these values along with the well-established equipartition analysis (Readhead 1994) to derive some general constraints on the properties of the emitting material. In particular, the energy of a synchrotron source with flux density, $F_p(\nu_p, t_p)$, can be expressed in terms of the equipartition energy density,

$$\frac{U}{U_{\text{eq}}} = \frac{1}{2}\epsilon_B \eta^{11}(1 + \frac{\epsilon_e}{\epsilon_B}\eta^{-17}),\tag{8.9}$$

where $\eta = \theta_s/\theta_{\text{eq}}$, the equipartition size is $\theta_{\text{eq}} \approx 120 d_{\text{Mpc}}^{-1/17} F_{p,\text{mJy}}^{8/17} \nu_{p,\text{GHz}}^{(-2\beta-35)/34}$ μas, $U_{\text{eq}} = 1.1 \times 10^{56} d_{\text{Mpc}}^2 F_{p,\text{mJy}}^4 \nu_{p,\text{GHz}}^{-7} \theta_{\text{eq},\mu\text{as}}^{-6}$ erg, and ϵ_e and ϵ_B are the fractions of energy in the electrons and magnetic fields, respectively. In equipartition $\epsilon_e = \epsilon_B = 1$ and the energy is minimized; a deviation from equipartition will increase the energy significantly.

For SN 2002cj we find $\theta_{\text{eq}}(t = 43-55\,\text{d}) \approx 30$ μas (i.e., $R_{\text{eq}} \approx 5\times10^{16}$ cm), which indicates an average expansion velocity, $v_{\text{eq}} \approx (0.35 - 0.45)c$. The equipartition energy is $U_{\text{eq}} \approx 8 \times 10^{47}$ erg, indicating a magnetic field strength of $B_{\text{eq}} \approx 0.2$ G.

We thank B. Paczynski, B. Schmidt, C. Wheeler and the referee for helpful comments. GRB and SN research at Caltech is supported in part by funds from NSF and NASA.

Table 8.1.   Radio Observations of Type Ib/c Supernovae in the Period 1999-2002

SN	IAUC	t_0 (UT)	t_{obs} (UT)	Δt (days)	Dist. (Mpc)	Detected?	$F_{8.46}$ (μJy)	$F_{4.86}$ (μJy)	$F_{1.43}$ (μJy)
1999ex	7310	Oct 25.6–Nov 9.5	Nov 18.02	8.5–23.4	54	No	±53[a]	±71	—
2000C	7348	1999 Dec 30–2000 Jan 8.3	Feb 4.04	26.7–36.0	59	Hint	—	187 ± 84	—
2000F	7353	1999 Dec 30.3–Jan 10.2	—	—	92	—	—	—	—
2000S	7384	<Feb 28[b]	—	—	94	—	—	—	—
2000cr	7443	Jun 21.2–25.9	Jul 3.04	7.1–11.8	54	No	±42[c]	—	—
2000de	7478	Jun 6–Aug 10.9	—	—	39	—	—	—	—
2000ds	7507	May 28–Oct 10.4	—	—	21	—	—	—	—
2000dt	7508	Sep 21–Oct 13.1	—	—	107	—	—	—	—
2000dv	7510	<Oct 17.1	—	—	63	—	—	—	—
2000ew	7530	May 8–Nov 28.5	2002 Jun 26	575–780	15	No	—	—	±67
2000fn	7546	Nov 1–19	Dec 29.22	40.2–58.2	72	No	±45	—	—
2001B	7555	2000 Dec 24.5–Jan 3.6	Feb 4.49	31.9–42.0	24	Yes	3500 ± 29	—	—
			2002 Oct 28.45	672.9–683		No	±40		
2001M	7568	Jan 3.4–21.4	Feb 4.45	14.1–32.1	56	No	±28	—	—
2001ai	7605	Mar 19.4–28.4	Mar 31.36	3–12	118	No	±43[f]	±49[f]	±49[f]
2001bb	7614	Apr 15.3–29.3	May 5.26	6–20	72	No	±50	±60	±120
2001ch	7637	2000 Nov 28.3–May 28.5	—	—	46	—	—	—	—
2001ci	7618	Apr 17.2–25.2	Jun 6.35	42.2–50.2	17	Yes	1450 ± 250	—	—
			2002 Jun 10.90	411.7–419.7		No	±150		
2001ef	7710	Aug 29–Sep 9.1	Sep 14.3	5.2–16.3	38	No	±115[g]	—	±88
			2002 Jun 18.0	281.9–293					
2001ej	7719	Sep 1–17.1	Sep 27.25	10.6–26.7	63	No	±44[h]	—	±85
			2002 Jun 14.0+18.0	269.9–469					
2001em	7722	Sep 10.3–15.3	—	—	91	—	—	—	—
2001eq(?)	7728	Aug 31.3–Sep 12.3	—	—	118	—	—	—	—
2001fw	7751	Oct 26–Nov 11.2	—	—	139	—	—	—	—
2001fx	7751	Oct 14.2–Nov 8.2	—	—	126	—	—	—	—
2001is	7782	Dec 14–23	2002 Jun 14.0+18.0	175–184	60	No	—	—	±70[i]
2002J	7800	Jan 9.4–21.4	Jun 18.0	147.6–159.6	58	No	—	—	±85
2002ap[j]	7810	Jan 28.5	Feb 1.03	3.5	7	Yes	374 ± 29	—	—
2002bl	7845	Feb 14–Mar 2.9	Mar 8.26	6–22	74	No	±50	±45	—
2002bm	7845	Jan 16–Mar 6.2	Jun 26.0	111.8–161	85	No	—	—	±66
2002cg	7877	Mar 28.5–Apr 13.5	Jun 26.0	73.5–89.5	150	No	—	—	±74[k]

Table 8.1

SN	IAUC	t_0 (UT)	t_{obs} (UT)	Δt (days)	Dist. (Mpc)	Detected?	$F_{8.46}$ (μJy)	$F_{4.86}$ (μJy)	$F_{1.43}$ (μJy)
2002cj	7882	Apr 9.5–21.5	Jun 3.19	42.7–54.7	106	Yes	112 ± 33	220 ± 37	240 ± 48
			Jun 18.0	57.5–69.5		Yes	—	—	408 ± 81
			Jun 26.0	65.5–77.5		Yes	—	—	300 ± 68
2002cp	7887	Apr 11.2–28.2	Jun 1.96	34.8–51.8	80	No	±31	±35	±36
2002cw		2001 Oct 1.2–May 16.5	—	—	71	—			
2002dg	7915	2002 May 5–May 31.3	Jul 9.95	39.7–66	215[l]	Hint	±33	92 ± 37	±46
2002dn(?)	7922	May 31.5–Jun 15.5	Jul 4.4	18.9–33.9	115	No	±39	±43	±72
2002dz	7935	2001 Nov 10.2–Jul 16.5	—	—	84	—			
2002ex	7964	Aug 19.3	—	—	180	—			
2002fh(?)	7971	Apr 9–May 9.3	—	—	1870	—			
2002ge(?)[m]	7987	2000 Oct 1.8–2002 Oct 7.9	Oct 12.0	~13	47	No	±160[n]	±130[o]	—
2002gy	7996	Oct 9.4–16.4	Oct 28.45	19.1–26.1	114	No	±36	±39	—
2002hf	8004	Oct 22.3–29.3	Nov 7.17	8.9–16.9	88	No	±40	—	—
2002hn	8009	Oct 21.5–30.5	Nov 7.42	7.9–16.9	82	No	±44	—	—
2002ho	8011	May 27–Nov 5.1	Nov 15.33	~56[p]	42	No	±47	±54	—
2002hy	8016	Oct 13.1–Nov 12.1	Nov 21.71	9.6–39.6	58	No	±74	—	—
2002hz	8017	Nov 2.2–12.2	2003 Jan 21.09	69.9–70.9	85	No	±29	—	—
2002ji	8025	Apr 10–Nov 30.8	Dec 5.67	38.4–59.4[q]	23	No	±43	—	—
2002jj	8026	Oct 1.4–24.4	Dec 15.28	51.9–74.9	66	No	±33	—	—
2002jp	8031	May 14.2–Nov 23.5	Dec 14.55	~33[r]	58	No	±38	—	—
2002jz	8037	2001 Dec 5–2002 Dec 23.3	2003 Jan 3.28	~32[s]	24	No	±25[t]	—	—

Note. — The columns are (left to right): (1) SN name, (2) IAU Circular number for the initial detection, (3) time of the SN explosion, with the range given between the most recent observation of the galaxy which did not show the SN and the epoch at which the SN was detected, (4) epoch of our VLA observations, (5) time delay between the SN explosion and the epoch of our observations, (6) distance to the galaxy (assuming $H_0 = 65$ km s^{-1} Mpc^{-1}), (7) indicates whether radio emission was detected, (8) flux density at 8.46 GHz, (9) flux density at 4.86 GHz, and (10) flux density at 1.43 GHz.

[a] uncertainties are quoted as 1σ rms; [b] nebular phase; [c] falls on top of galaxy substructure, $< 10\sigma$; [d] Ia or Ic (IAUC 7574); [e] before maximum (IAUC 7574); [f] falls on top of galaxy with complex substructure, $< 4\sigma$; [g] falls on top of galaxy, $< 5\sigma$; [h] falls on top of galaxy, $< 35\sigma$; [i] on top of galaxy substructure, $< 3\sigma$; [j] See Berger et al. (2002b); [k] on top of galaxy, $< 10\sigma$; [l] IAUC 7990; [m] Type Ic similar to SN 1994I (IAUC 7574); [n] on top of galaxy, $< 10\sigma$; [o] on top of galaxy, $< 3\sigma$; [p] within a few weeks past maximum brightness (IAUC 8014); [q] 3 – 6 weeks past maximum light (IAUC 8026); [r] two weeks past maximum light (IAUC 8031); [s] ten days past maximum (IAUC 8037); [t] on top of galaxy substructure, $< 3\sigma$.

Table 8.2.   Ejecta and Progenitor Properties of Type Ib/c Supernovae Detected in the Radio

SN	E_K $(10^{51}$ erg)	M_{ej} $(M_\odot)$	$M_{^{56}Ni}$ $(M_\odot)$	M_{CO} $(M_\odot)$	M_{prog} $(M_\odot)$	Ref.
1984L	20	50	0.2	—	—	1
	—	10	—	—	—	2
	—	—	0.1	—	$20-30$	3
1994I	$1-1.4$	$0.9-1.3$	0.07	<1.5	<15	4
	—	0.9	0.07	1.35	—	5
	1	—	—	—	—	6
1998bw	30	—	0.7	13.8	40	7
	50	10	0.4	13.8	40	8
	22	—	0.5	6.5	—	9
	2	2	0.2	—	—	10
	—	—	$0.5-0.9^a$	—	—	11
2002ap	$4-10$	$2.5-5$	0.07	5	$20-25$	12

Note. — The columns are (left to right): (1) Ejecta kinetic energy, (2) ejecta mass, (3) ^{56}Ni mass, (4) estimated mass of the CO core, (5) estimated mass of the progenitor, and (6) references. Data are not available for the SNe detected in this survey. ^{a} These authors use the models of Iwamoto et al. (1998) and Woosley et al. (1999) as input; they assert that observations in the nebular phase exclude the low ^{56}Ni mass inferred by Höflich et al. (1999).

References. — (1) Baron et al. (1993); (2) Swartz & Wheeler (1991); (3) Schlegel & Kirshner (1989); (4) Young et al. (1995); (5) Iwamoto et al. (1994); (6) Millard et al. (1999); (7) Iwamoto et al. (1998); (8) Nakamura et al. (2001); (9) Woosley et al. (1999); (10) Höflich et al. (1999); (11) Sollerman et al. (2000); (12) Mazzali et al. (2002)

Table 8.3.   Best-Fit Models for the Supernova and γ-Ray Burst Luminosity Distributions

Population	Model	Parameters	$\log(\mathcal{L})/$dof
SN	Gaussian	(26.1, 1.0)	22.5/36
SN	D. Powerlaw	(-29.0, 25.4)	22.4/36
SN	Flat	(20.0, 29.1)	22.3/36
GRB	Gaussian	(31.0, 0.8)	38.8/33
GRB	D. Powerlaw	(-22.5, 29.6)	48.9/33
GRB	Flat	(29.6, 32.4)	37.7/33
SN+GRB	Gaussian	(28.4, 2.4)	130.3/71
SN+GRB	D. Powerlaw	(-10.4, 25.3)	123.8/71
SN+GRB	I. Powerlaw	(3.7, 32.3)	103.7/71
SN+GRB	Flat	(22.6, 32.4)	104.9/71

Note. — The columns are (left to right): (1) Data set, (2) population distribution function, (3) best-fit parameters, and (4) log likelihood. A detailed explanation of the models and the fitting procedure is provided in §8.4.

Part III

The Multi-Wavelength Properties of Gamma-Ray Burst Host Galaxies

CHAPTER 9

The Faint Optical Afterglow and Host Galaxy of GRB 020124: Implications for the Nature of Dark Gamma-Ray Bursts[†]

E. Berger[a], S. R. Kulkarni[a], J. S. Bloom[a], P. A. Price[a,b], D. W. Fox[a], D. A. Frail[a,c], T. S. Axelrod[b], R. A. Chevalier[d], E. Colbert[e], E. Costa[f], S. G. Djorgovski[a], F. Frontera[g,h], T. J. Galama[a], J. P. Halpern[i], F. A. Harrison[a], J. Holtzman[j], K. Hurley[k], R. A. Kimble[l], P. J. McCarthy[m], L. Piro[f], D. Reichart[a], G. R. Ricker[n], R. Sari[o], B. P. Schmidt[b], J. C. Wheeler[p], R. Vanderspek[n], & S. A. Yost[a]

[a]Department of Astronomy, 105-24 California Institute of Technology, Pasadena, CA 91125, USA

[b]Research School of Astronomy & Astrophysics, Mount Stromlo Observatory, via Cotter Rd., Weston Creek 2611, Australia

[c]National Radio Astronomy Observatory, Socorro, NM 87801

[d]Department of Astronomy, University of Virginia, P.O. Box 3818, Charlottesville, VA 22903-0818

[e]Department of Physics and Astronomy, Johns Hopkins University, 3400 N. Charles St., Baltimore, MD 21218

[f]Istituto Astrofisica Spaziale, C.N.R., Area di Tor Vergata, Via Fosso del Cavaliere 100, 00133 Roma, Italy

[g]Istituto Astrofisica Spaziale and Fisica Cosmica, C.N.R., Via Gobetti, 101, 40129 Bologna, Italy

[h]Physics Department, University of Ferrara, Via Paradiso, 12, 44100 Ferrara, Italy

[i]Columbia Astrophysics Laboratory, Columbia University, 550 West 120th Street, New York, NY 10027

[j]Department of Astronomy, MSC 4500, New Mexico State University, P.O. Box 30001, Las Cruces, NM 88003

[k]University of California at Berkeley, Space Sciences Laboratory, Berkeley, CA 94720-7450

[l]Laboratory for Astronomy and Solar Physics, NASA Goddard Space Flight Center, Code 681, Greenbelt, MD 20771

[m]Carnegie Observatories, 813 Santa Barbara Street, Pasadena, CA 91101

[n]Center for Space Research, Massachusetts Institute of Technology, 70 Vassar Street, Cambridge, MA 02139-4307

[o]Theoretical Astrophysics 130-33, California Institute of Technology, Pasadena, CA 91125

[p]Astronomy Department, University of Texas, Austin, TX 78712

[†] A version of this chapter was published in *The Astrophysical Journal*, vol. 590, 379–385, (2003).

Abstract

We present ground-based optical observations of GRB 020124 starting 1.6 hours after the burst, as well as subsequent Very Large Array (VLA) and *Hubble Space Telescope* (HST) observations. The optical afterglow of GRB 020124 is one of the faintest afterglows detected to date, and it exhibits a relatively rapid decay, $F_\nu \propto t^{-1.60\pm0.04}$, followed by further steepening. In addition, a weak radio source was found coincident with the optical afterglow. The HST observations reveal that a positionally coincident host galaxy must be the faintest host to date, $R \gtrsim 29.5$ mag. The afterglow observations can be explained by several models requiring little or no extinction within the host galaxy, $A_V^{host} \approx 0 - 0.9$ mag. These observations have significant implications for the interpretation of the so-called dark bursts (bursts for which no optical afterglow is detected), which are usually attributed to dust extinction within the host galaxy. The faintness and relatively rapid decay of the afterglow of GRB 020124, combined with the low inferred extinction indicate that some dark bursts are intrinsically dim and not dust obscured. Thus, the diversity in the underlying properties of optical afterglows must be observationally determined before substantive inferences can be drawn from the statistics of dark bursts.

SECTION 9.1
Introduction

One of the main observational results stemming from five years of γ-ray burst (GRB) follow-ups at optical wavelengths is that about 60% of well-localized GRBs lack a detected optical afterglow, ("dark bursts"; Taylor et al. 2000; Fynbo et al. 2001; Lazzati et al. 2002; Reichart & Yost 2001. In some cases, a non-detection of the optical afterglow could simply be due to a failure to image quickly and/or deeply enough. However, there are two GRBs for which there is strong evidence that the optical emission should have been detected, based on an extrapolation of the radio and X-ray emission (Djorgovski et al. 2001a; Piro et al. 2002). One interpretation in these two cases is that the optical light was extinguished by dust, either within the immediate environment of the burst or elsewhere along the line of sight (e.g., Groot et al. 1998. An alternative explanation is a high redshift, leading to absorption of the optical light in the Lyα forest. However, the redshifts of the underlying host galaxies of these GRBs are of order unity (Djorgovski et al. 2001a; Piro et al. 2002).

Several authors have recently argued that a large fraction of the dark bursts are due to dust extinction within the local environment of the bursts (e.g., Lazzati et al. 2002; Reichart & Price 2002; Reichart & Yost 2001), but other scenarios have also been suggested (e.g., Lazzati et al. 2002). Moreover, it has been noted that regardless of the location of extinction within the host galaxy, the fraction of dark bursts is a useful upper limit on the fraction of obscured star formation (Kulkarni et al. 2000; Djorgovski et al. 2001b; Ramirez-Ruiz et al. 2002; Reichart & Price 2002).

However, from an observational point of view, we must have a clear understanding of the diversity of afterglow properties before extracting astrophysically interesting inferences from dark bursts. For example, afterglows which are intrinsically faint or fade rapidly (relative to the detected population) would certainly bias the determination of the fraction of truly obscured bursts. In this vein, Fynbo et al. (2001), noting the faint optical afterglow of GRB 000630, argue that some dark bursts are due to a failure to image deeply and/or quickly enough, rather than dust extinction. Observations of the faint afterglow of GRB 980613 (Hjorth et al. 2002) point to the same conclusion.

Here we present optical and radio observations of GRB 020124, an afterglow that would have been classified dark had it not been for rapid and deep searches. Furthermore, GRB 020124 is an example of an afterglow, which is dim due to the combination of intrinsic faintness and a relatively fast decline, and not strong extinction.

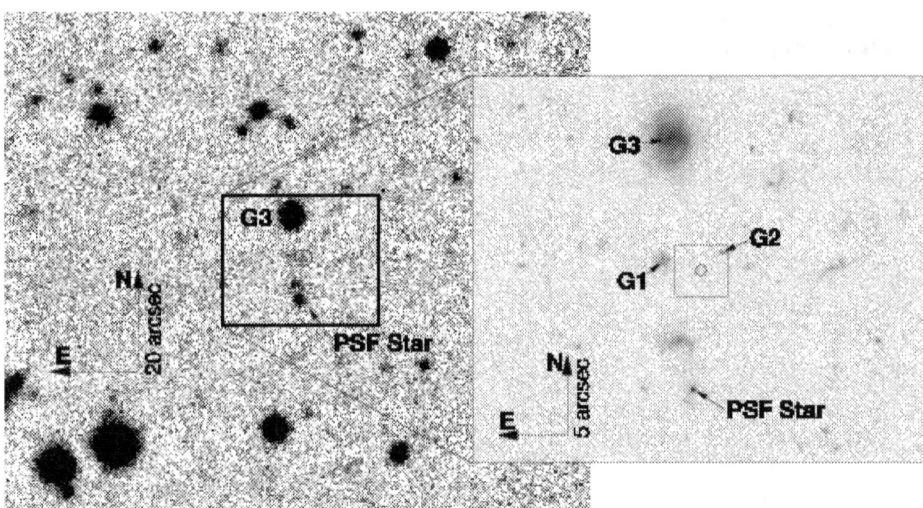

Figure 9.1: Palomar 200-inch (left) and HST epoch 1 (inset) images of the field of GRB 020124. The OT is circled in both images. The OT was of comparable brightness to G1 at the epoch of the P 200 image and significantly fainter than G1 three weeks later. The box overlaying the inset shows the portion of the HST images depicted in Figure 9.2. Relevant sources described in the text are noted. The HST image is shown with logarithmic scaling to highlight the features of nearby galaxies.

SECTION 9.2

Observations

9.2.1   Ground-Based Observations

GRB 020124, localized by the HETE-II satellite on 2002, Jan 24.44531 UT, had a duration of ~ 70 s and a fluence $(6 - 400$ keV) of 3×10^{-6} erg cm^{-2} (Ricker et al. 2002). Eight minutes after receiving the coordinates[1] we observed the error box with the dual-band (B_M, R_M) MACHO imager mounted on the robotic 50-in telescope at the Mount Stromlo Observatory (MSO). We also observed the error box with the Wide-Field Imager on the 40-in telescope at Siding Spring Observatory (SSO). We were unable to identify a transient source within the large error box (Price et al. 2002c).

We subsequently observed the error box refined by the Inter-Planetary Network (Hurley et al. 2002) with the Palomar 48-in Oschin Schmidt using the unfiltered NEAT imager. PSF-matched image subtraction (Alard 2000) between the MACHO and NEAT images revealed a fading source (Price et al. 2002b), which was $R \approx 18$ mag at the epoch of our first observations, and not present in the Digitized Sky Survey. Two nights later we observed the afterglow using the Jacobs CAMera (JCAM) mounted at the East arm focus of the Palomar 200-in telescope (Bloom 2002). The position of the fading source is α(J2000)=9$^{\rm h}$32$^{\rm m}$50.78$^{\rm s}$, δ(J2000)=$-11°31'10.6''$, with an uncertainty of about 0.4 arcsec in each coordinate (Figure 9.1).

Using the Very Large Array (VLA[2]) we observed the fading source at 8.46 and 22.5 GHz (see Table 9.3). We detect a faint source, possibly fading, at 8.46 GHz located at α(J2000)=9$^{\rm h}$32$^{\rm m}$50.81$^{\rm s}$, δ(J2000)= $-11°31'10.6''$, with an uncertainty of about 0.1 arcsec in each coordinate. Given the positional coincidence between the fading optical source and radio detection we suggest this source to be the afterglow of GRB 020124.

The optical images were bias-subtracted and flat-fielded in the standard manner. To extract the

---

[1] This corresponds to 1.6 hours after the burst detection.

[2] The VLA is operated by the National Radio Astronomy Observatory, a facility of the National Science Foundation operated under cooperative agreement by Associated Universities, Inc.

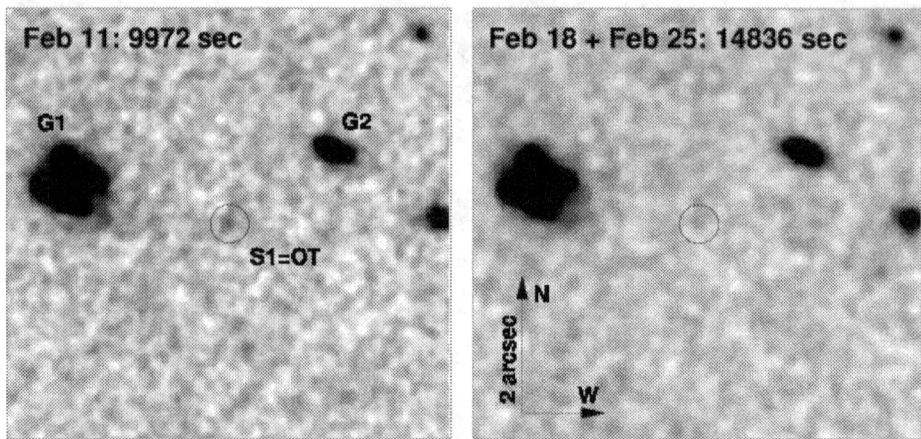

Figure 9.2: The faint optical transient (OT) of GRB 020124 as viewed using HST/STIS. Shown are the summed, smoothed images from epoch 1 (left) and epochs 2+3 (right). The grey-scales have been matched such that a given flux is represented by the same shade in each image. The circle is centered at the same sky position in both images. Clearly, the source S1, identified with the position of the afterglow of GRB 020124 has faded.

photometry we weighted the aperture with a Gaussian equivalent to the seeing disk ("weighted-aperture photometry"), using IRAF/wphot. The photometric zero-points were set through photometry of calibrated field stars (Henden 2002) with magnitudes transformed to the appropriate system (Bessell & Germany 1999; Smith et al. 2002b). The photometry is summarized in Table 9.1.

9.2.2 *Hubble Space Telescope* **Observations**

We observed the afterglow with the *Hubble Space Telescope* (HST) using the Space Telescope Imaging Spectrograph (STIS) on 2002 Feb. 11.09, 18.30, and 25.71 UT Bloom et al. (2002c), as part of our HST Cycle 10 program (GO-9180, PI: Kulkarni). The HST observations consisted of 750–850 sec exposures. The HST data were retrieved after "On-The-Fly" pre-processing. Using IRAF we drizzled (Fruchter & Hook 2002) each image onto a grid with pixels smaller than the original by a factor of two and using pixfrac of 0.7.

We found an astrometric tie between the HST and JCAM images using IRAF/geomap with nine suitable astrometric tie objects in common between the images. The rms of the resultant mapping is 133 mas (RA) and 124 mas (Dec). Using this mapping and IRAF/geoxytrans we transfered the afterglow position on the JCAM image to the HST images. The rms of the transformation is 604 mas (RA) and 596 mas (Dec), and is dominated by the uncertainty in the JCAM position.

The source "S1" (Figure 9.2) coincides with the afterglow position within the astrometric uncertainty. We performed differential photometry at the position of S1 by registering the images of epochs 1 and 2 using a cross-correlation of a field of size 10 arcsec centered on S1 (using IRAF/crosscor and shiftfind). We used IRAF/center and the FWHM of a relatively bright point source ("PSF star"; Figure 9.1) to fix the position of S1 in each of the final images, and to determine the uncertainty in the position.

We photometered the source (and the PSF star) in epoch 1 using IRAF/phot, in a 3.4 pix (86 mas) drizzled aperature radius. The small radius was chosen to maximize the signal-to-noise of the detection of the faint point source although, as found using the STIS instrument manual and confirmed with the PSF star, this radius encircles only $\sim 55\%$ of the light of a point source. A corresponding correction was applied to the fluxes found in this aperature; we estimate a 0.1 mag sytematic uncertainty due to this correction. Using IRAF/synphot, and assuming a source spectrum of $f_\lambda \propto \lambda^{-0.5}$ (see below), we find that the source was $R = 28.68^{-0.20}_{+0.25}$ mag at the time of epoch 1. A bluer spectrum would result in an even fainter R-band magnitude, by as much as 0.25 mag for $f_\lambda \propto \lambda^{-2.5}$. More importantly, a redder spectrum would have little effect at R-band, with an increase of < 0.05 mag. The photometry of the three epochs is summarized in Table 9.2. Note that this more careful analysis supersedes our preliminary report (Bloom et al. 2002c).

There are no obvious persistent sources within 1.75 arcsec of the OT down to $R \approx 29.5$ mag. To date, all of the GRBs localized to sub-arcsecond accuracy have viable hosts brighter than this level within ~ 1.3 arcsec of the OT position (Bloom et al. 2002a). The faintest host to date is that of GRB 990510, $R \sim 28.5$ mag (z=1.619; Vreeswijk et al. 2001b). Thus, the host of GRB 020124 may be at a somewhat higher redshift; however, $z \lesssim 4.5$ since the afterglow was detected in the B_M filter.

SECTION 9.3

Modeling of the Optical Data

In Figure 9.3 we plot the optical light curves of GRB 020124, including a correction for Galactic extinction, $E(B - V) = 0.052$ mag (Schlegel et al. 1998). The optical light curves are usually modeled as $F_\nu(t, \nu) = F_{\nu,0}(t/t_0)^\alpha (\nu/\nu_0)^\beta$. However, as can be seen in Figure 9.3, the R-band light curve cannot be described by a single power law. Restricting the fit to $t < 2$ days we obtain ($\chi^2_{\min} = 15$ for 14 degrees of freedom) $\alpha_1 = -1.60 \pm 0.04$, $\beta = -1.43 \pm 0.14$, and $F_{\nu,0} = 2.96 \pm 0.25$ μJy; here $F_{\nu,0}$ is defined at the effective frequency of the R_M filter and $t = 1$ day. For $t > 2$ days we get $\alpha_2 = -1.9^{+1.0}_{-2.0}$. The uncertainty in α_2 is large because it is effectively constrained by only two data points. However, if we make the additional requirement that the fits to the ground-based data and the HST data intersect at $t > 2$ days, we find that $\alpha_2 = -1.9^{+0.1}_{-2.0}$, and the steepening is therefore significant at the 2.5σ level.

To account for the steepening we modify the model for the R-band light curve to:

$$F_\nu(t, \nu) = F_{\nu,0}(\nu/\nu_0)^\beta [(t/t_b)^{\alpha_1 n} + (t/t_b)^{\alpha_2 n}]^{1/n}, \tag{9.1}$$

where, α_1 is the asymptotic index for $t \ll t_b$, α_2 is the asymptotic index for $t \gg t_b$, $n < 0$ provides a smooth joining of the two asymptotic segments, and t_b is the time at which the asymptotic segments intersect. We retain the simple model for the R_M and B_M light curves since they are restricted to $t \lesssim 0.13$ days (i.e., well before the observed steepening).

We investigate two alternatives for the observed steepening in the framework of the afterglow synchrotron model (e.g., Sari et al. 1998), namely (i) a cooling break (§9.3.1), and (ii) a jet break (§9.3.2). In this framework, α_1, α_2, and β are related to each other through the index (p) of the electron energy distribution, $N(\gamma) \propto \gamma^{-p}$ (for $\gamma > \gamma_{\min}$). The relations for the models discussed below, as well as the resulting closure relations, $\alpha_1 + b\beta + c = 0$, are summarized in Table 9.4.

9.3.1 Cooling Break

The observed steepening, $\Delta\alpha \equiv \alpha_2 - \alpha_1 \approx -0.3$, can be due to the passage of the synchrotron cooling frequency, ν_c, through the R-band[3] . This has been suggested, for example, in the afterglow of GRB 971214, at $t \sim 0.6$ days (Wijers & Galama 1999). If the steepening is due to ν_c, this rules out models in which the ejecta expand into a circumburst medium with $\rho \propto r^{-2}$ (hereafter, Wind), because in this model ν_c increases with time ($\propto t^{1/2}$; Chevalier & Li 1999), and one expects $\Delta\alpha = 0.25$.

There are two remaining models to consider in this case: (i) spherical expansion into a circumburst medium with constant density (hereafter, ISM$_B$; Sari et al. 1998), and (ii) a jet with $\theta_{\rm jet} < \Gamma^{-1}_{t\sim0.06\,{\rm d}}$ (i.e.,

---

[3] We note that while the passage of ν_c through the R-band will also change the spectrum of the afterglow by $\delta\beta = -0.5$ (i.e., the afterglow would become somewhat redder), this has little effect on the conversion of the STIS count-rate to R-band magnitudes (see §9.2.2). We therefore use the same source magnitudes listed in Table 9.2, along with the relevant systematic uncertainties.

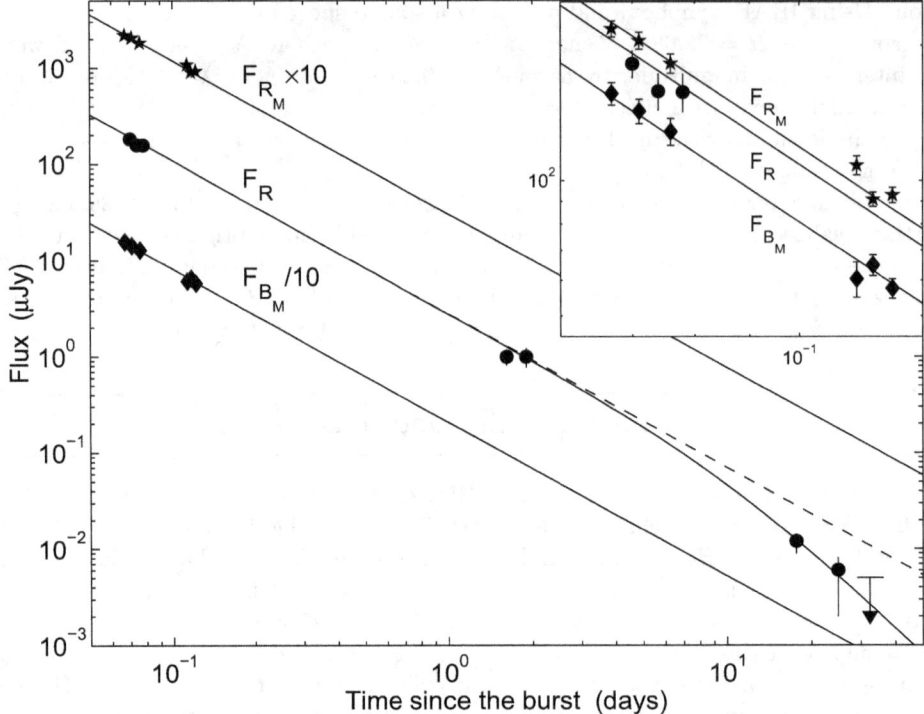

Figure 9.3: Optical light curves of GRB 020124 (top to bottom: R_M, R, and B_M), corrected for Galactic extinction, $E(B-V) = 0.052$ mag (Schlegel et al. 1998). The solid lines are a representative jet model (ISM/Wind$_R$; see Table 9.4), while the dashed line is an extrapolation of the early evolution without a break. With no break in the R-band light curve, the predicted magnitude at the epoch of the first HST observation exceeds the measured values by 5σ. The flux measured in the last HST epoch is plotted as a 2σ upper limit.

a jet break prior to the first observation at $t \approx 0.06$ days; hereafter, Jet$_B$). The subscript B indicates that ν_c is blueward of the optical bands initially. In both models we use Eqn. 9.1 for the R-band light curve, with t_b defined as the time at which $\nu_c = \nu_R$, and $\alpha_2 \equiv \alpha_1 - 1/4$.

We find that in the ISM$_B$ model $t_c \approx 0.4$ days, while in the Jet$_B$ model $t_c \approx 0.65$ days. Moreover, in both models the closure relations can only be satisfied by including a contribution from dust extinction within the host galaxy, $A_V^{\rm host}$. We estimate the required extinction using the parametric extinction curves of Cardelli et al. (1989) and Fitzpatrick & Massa (1988), along with the interpolation calculated by Reichart (2001). Since the redshift of GRB 020124 is not known we assume $z = 0.3, 1, 3$, which spans the range of typical redshifts for the long-duration GRBs. The inferred values of $A_V^{\rm host}$ are summarized in Table 9.4, and range from 0.2 to 0.9 mag.

9.3.2 Jet Break

An alternative explanation for the steepening is a jet expanding into: (i) an ISM medium with ν_c blueward of the optical bands (J-ISM$_B$), (ii) a Wind medium with ν_c blueward of the optical bands (J-Wind$_B$), and (iii) an ISM or Wind medium with ν_c redward of the optical bands (J-ISM/Wind$_R$). We note that the J-ISM$_B$ model is different than the ISM$_B$ model (§9.3.1) since previously it was implicitly defined such that the jet break is later than the last observation. In these models, $t_b \equiv t_{\rm jet}$ is the time at which $\Gamma(t_{\rm jet}) \approx \theta_{\rm jet}^{-1}$.

From the closure values we note that the J-ISM/Wind$_R$ requires no extinction within the host galaxy,

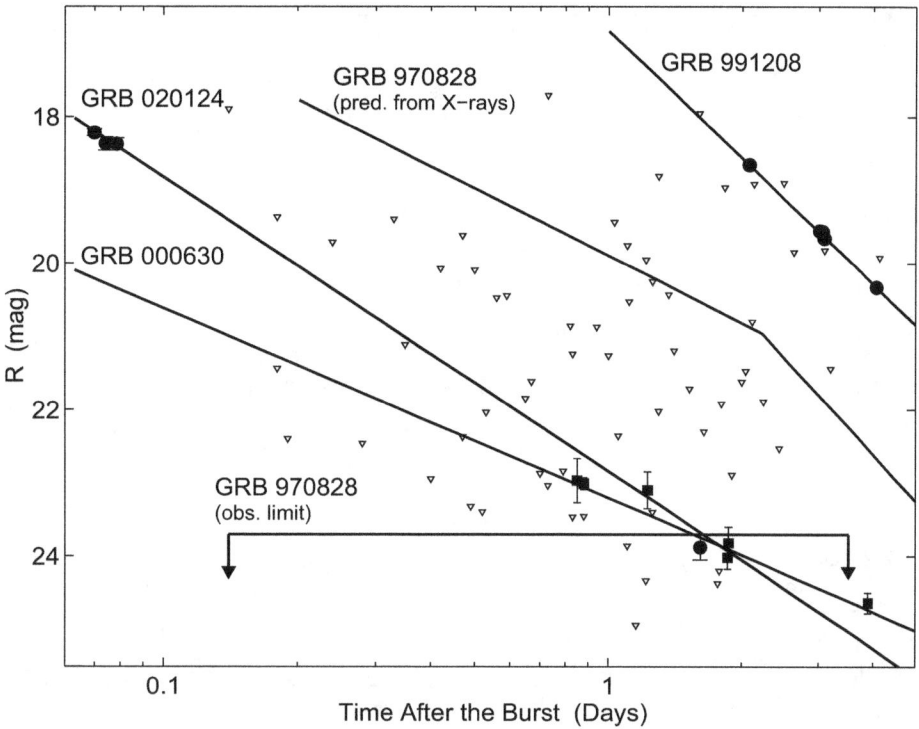

Figure 9.4: *R*-band upper limits from searches of well-localized GRBs, corrected for Galactic extinction. The limits up to GRB 000630 are taken from Fynbo et al. (2001), while subsequent limits are from the GRB Coordinates Network. Also shown are the light curves of the GRB 020124, GRB 000630, the bright GRB 991208 Castro-Tirado et al. (2001), and GRB 970828 (the de-reddened light curve is based on the radio and X-ray data; Djorgovski et al. 2001a). Only about 30% of the searches yielded limits that are fainter than the afterglow of GRB 020124. A similar fraction was found by Fynbo et al. (2001) based on the afterglow of GRB 000630.

while the J-ISM$_B$ and J-Wind$_B$ models require values of about 0.05 to 0.3 mag.

We find $t_{jet} \sim 10 - 20$ days, corresponding to $\theta_{jet} \sim 10°$. Using the measured fluence (§9.2.1) we estimate the beaming-corrected γ-ray energy, $E_\gamma \approx 5 \times 10^{50}\, n_1^{1/4}$ erg, assuming a circumburst density $n_1 = 1$ cm^{-3} and $z = 1$ (E_γ is a weak function of z). This value is in good agreement with the distribution of E_γ for long-duration GRBs (Frail et al. 2001).

SECTION 9.4

Discussion and Conclusions

Regardless of the specific model for the afterglow emission, the main conclusion of §9.3 is that the optical afterglow of GRB 020124 suffered little or no dust extinction. Still, this afterglow would have been missed by typical searches undertaken even as early as 12 hours after the GRB event. As shown in Figure 9.4, about 70% of the searches conducted to date would have failed to detect an optical afterglow like that of GRB 020124.

This is simply because the afterglow of GRB 020124 was faint and exhibited relatively rapid decay. From Figure 9.5 we note that GRB 020124 is one of the faintest afterglows detected to date (normalized to $t = 1$ day), and while it is not an excessively rapid fader, it is in the top 30% in this category.

Thus, the afterglow of GRB 020124, along with that of GRB 000630 (Fynbo et al. 2001; Figure 9.5)

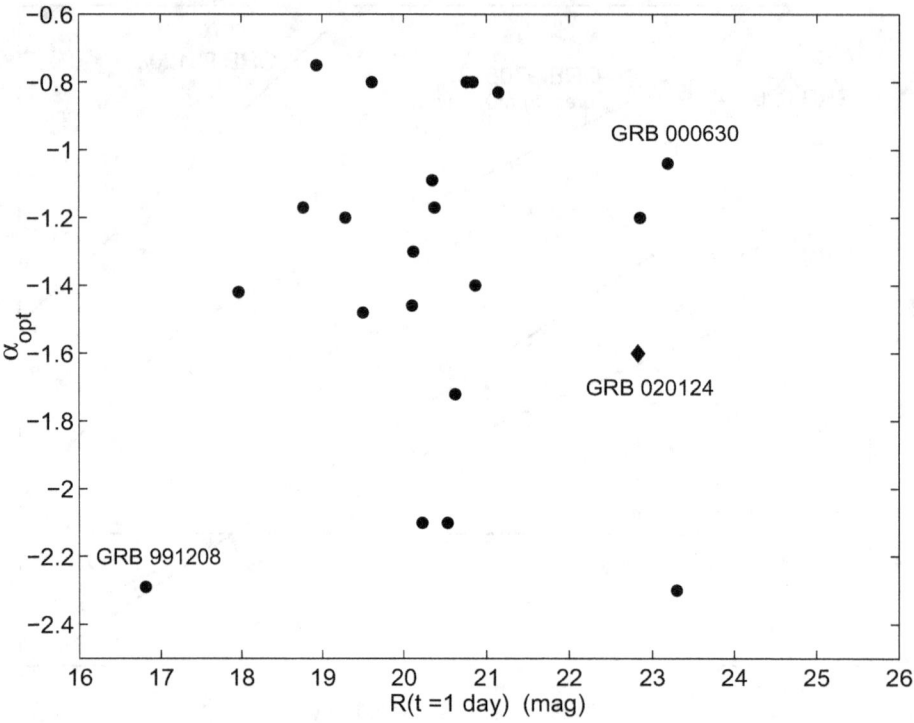

Figure 9.5: Temporal decay index, α_{opt} ($F_\nu \propto t^\alpha$), plotted against the R-band magnitude at $t = 1$ day for several optical afterglows. We chose a fiducial time of 1 day since, with the exception of GRB 010222, all the observations are before the jet break. While the majority of optical afterglows cluster around $R(t = 1\,d) \sim 20$ mag, GRB 020124 is one of the four faintest afterglows detected to date, and one of the six most rapid faders.

and GRB 980613 Hjorth et al. (2002), indicates that there is a wide diversity in the brightness and decay rates of optical afterglows. In fact, the brightness distribution spans a factor of about 400, while the decay index varies by more than a factor of three. Coupled with the low dust extinction in the afterglow of GRB 020124, this indicates that some dark bursts may simply be dim, and not dust obscured.

   Given this wide diversity in the brightness of optical afterglows, it is important to establish directly that an afterglow is dust obscured. This has only been done in a few cases (§9.1). Therefore, while *statistical* analyses (e.g., Reichart & Yost 2001) point to extinction as the underlying reason for some fraction of dark bursts, it is clear that observationally the issue of dark bursts is not settled, and the observational biases have not been traced fully (see also Fynbo et al. 2001).

   Since progress in our understanding of dark bursts will benefit from observations, we need consistent, rapid follow-up of a large number of bursts to constrain the underlying distribution, as well as complementary techniques which can directly measure material along the line of sight. This includes X-ray observations which allow us to measure the column density to the burst (Galama & Wijers 2001), and thus infer the type of environment, and potential extinction level. Along the same line, radio observations allow us to infer the synchrotron self-absorption frequency, which is sensitive to the ambient density (e.g., Sari & Esin 2001); the detection of radio emission, as in the case of GRB 020124, implies a density $n \lesssim 10^2$ cm^{-3}. Finally, prompt optical observations, as we have carried out in this case, may uncover a larger fraction of the dim optical afterglows, and provide a better constraint on the fraction of truly obscured bursts.

J. S. B. is a Fannie and John Hertz Foundation Fellow. F. A. H. acknowledges support from a Presidential Early Career award. S. R. K. and S. G. D. thank the NSF for support. R. S. is grateful for support from a NASA ATP grant. R. S. and T. J. G. acknowledge support from the Sherman Fairchild Foundation. J. C. W. acknowledges support from NASA grant NAG59302. K. H. is grateful for Ulysses support under JPL contract 958056 and for IPN support under NASA grants FDNAG 5-11451 and NAG 5-17100. Support for Proposal HST-GO-09180.01-A was provided by NASA through a grant from the Space Telescope Science Institute, which is operated by the Association of Universities for Research in Astronomy, Inc., under NASA contract NAS5-26555. We thank the anonymous referee for helpful comments.

Table 9.1.   Ground-Based Optical Observations of GRB 020124

UT	Telescope	Band	Magnitude
Jan 24.51204	MSO 50	R_M	17.918 ± 0.041
Jan 24.51204	MSO 50	B_M	18.628 ± 0.057
Jan 24.51516	SSO 40	R	18.219 ± 0.046
Jan 24.51655	MSO 50	R_M	17.984 ± 0.044
Jan 24.51655	MSO 50	B_M	18.727 ± 0.063
Jan 24.51938	SSO 40	R	18.371 ± 0.091
Jan 24.52106	MSO 50	R_M	18.111 ± 0.049
Jan 24.52106	MSO 50	B_M	18.842 ± 0.069
Jan 24.52373	SSO 40	R	18.376 ± 0.082
Jan 24.55791	MSO 50	R_M	18.678 ± 0.048
Jan 24.55791	MSO 50	B_M	19.661 ± 0.090
Jan 24.56243	MSO 50	R_M	18.867 ± 0.036
Jan 24.56243	MSO 50	B_M	19.584 ± 0.053
Jan 24.56696	MSO 50	R_M	18.843 ± 0.039
Jan 24.56696	MSO 50	B_M	19.714 ± 0.050
Jan 26.34100	P 200	r'	24.398 ± 0.228

Note. — The columns are (left to right), (1) UT date of each observation, (2) telescope (MSO 50: Mt. Stromlo Observatory 50-in; SSO 40: Siding Spring Observatory 40-in; P 200: Palomar Observatory 200-in), (3) observing band, and (4) magnitudes and uncertainties. The observed magnitudes are not corrected for Galactic extinction.

Table 9.2.   HST/STIS Observations of GRB 020124

Epoch (UT)	Band	Exp. Time (ksec)	Flux (e^- s^{-1})	S/N	Magnitude
Feb 11.09	50 CCD/Clear	10.0	0.0814 ± 0.0169	4.82	$R = 28.68^{+0.25}_{-0.20}$
Feb 18.30	50 CCD/Clear	7.4	0.0443 ± 0.0189	2.34	$R = 29.35^{+0.60}_{-0.39}$
Feb 25.71	50 CCD/Clear	7.5	0.0362 ± 0.0183	1.98	$R = 29.56^{+0.76}_{-0.44}$
Feb 18.30+25.71	50 CCD/Clear	14.9	0.0398 ± 0.0137	2.91	$R = 29.46^{+0.46}_{-0.32}$

Note. — The columns are (left to right), (1) UT date of each observation, (2) STIS CCD mode, (3) exposure time, (4) flux and uncertainty, (5) significance, and (6) R magnitude and uncertainty. The total number of counts was converted to the R-band assuming the observed color of the OT, $f_\lambda \propto \lambda^{-0.5}$ (§9.2.2). The R-band errors reflect only the statistical uncertainty. Choosing a wide range of assumed colors for the afterglow ($\alpha_\lambda = -2.5$ to 0.5) gives $+0.25$, -0.05 mag. Thus, the afterglow could not have been much brighter in R-band than reported in epoch 1. We include this color uncertainty in the analysis (§9.3), and in Figure 9.3, choosing half of the range as the rms of the systematic color uncertainty. In addition, we also include in the analysis the estimated uncertainty from the aperture correction (0.1 mag; §9.2.2). For epochs 2 and 3, the 3σ upper limits are: $R = 29.09$ and $R = 29.13$ mag, respectively. The observed magnitudes are not corrected for Galactic extinction.

Table 9.3.   VLA Radio Observations of GRB 020124

Epoch (UT)	ν_0 (GHz)	Flux Density (μJy)
Jan 26.22	8.46	84 ± 30
Jan 26.25	22.5	-60 ± 100
Jan 27.22	8.46	45 ± 25
Feb 1.40	8.46	49 ± 17
Jan 26.22-Feb 1.40	8.46	48 ± 13

Note. — The columns are (left to right), (1) UT date of each observation, (2) observing frequency, and (3) flux density at the position of the radio transient with the rms noise calculated from each image. The last row gives the flux density at 8.46 GHz from the co-added map.

Table 9.4.   Afterglow Models for GRB 020124

Model	α_1	α_2	β	(b, c)	Closure	p	$A_V^{\rm host}$ (mag)
ISM$_{\rm B}$	$-\frac{3(p-1)}{4}$	$-\frac{3p}{4} + \frac{1}{2}$	$-\frac{p-1}{2}$	$(-3/2, 0)$	0.52 ± 0.28	3.17 ± 0.05	$(0.35, 0.18, 0.10)$
Jet$_{\rm B}$	$-p$	$-p$	$-\frac{p-1}{2}$	$(-2, 1)$	2.23 ± 0.36	1.63 ± 0.04	$(0.89, 0.50, 0.22)$
J-ISM$_{\rm B}$	$-\frac{3(p-1)}{4}$	$-p$	$-\frac{p-1}{2}$	$(-3/2, 0)$	0.52 ± 0.28	3.17 ± 0.05	$(0.30, 0.10, 0.05)$
J-Wind$_{\rm B}$	$-\frac{3p-1}{4}$	$-p$	$-\frac{p-1}{2}$	$(-3/2, 1/2)$	1.02 ± 0.28	2.51 ± 0.05	$(0.30, 0.16, 0.08)$
J-ISM/Wind$_{\rm R}$	$-\frac{3p-2}{4}$	$-p$	$-\frac{p}{2}$	$(-3/2, -1/2)$	0.02 ± 0.28	2.84 ± 0.05	—

Note. — The columns are (left to right), (1) Afterglow model (ISM: r^0 circumburst medium; Wind: r^{-2} circumburst medium; Jet: collimated eject with opening angle $\theta_{\rm jet}$; a subscript B indicates $\nu_c < \nu_{\rm opt}$, and a subscript R indicates $\nu_c > \nu_{\rm opt}$), (2) α_1 as a function of p, (3) α_2 as a function of p, (4) β as a function of p, (5) closure relations ($\alpha + b\beta + c = 0$), (6) resulting closure values from the observed values of α_1 and β, (7) inferred value of p from the measured value of α_1, and (8) the required extinction in the frame of the host galaxy for closure values of zero ($z = 0.3, 1, 3$); typical uncertainties are ± 0.05 mag. The top two models apply to the case when the observed steepening in the light curves is due to the passage of ν_c through the R-band, while the bottom three apply to the case when the steepening is due to a jet.

Table 9.5.   Limits on Optical Afterglow Magnitudes for Bursts Localized in 2000–2002

GRB	Epoch (days)	R-limit (mag)	Reference
GRB 000801	1.77	24.5	GCN 767
GRB 000812	4.14	20.8	GCN 771
GRB 000830	0.99	24.5	GCN 788
GRB 001025	1.21	24.5	GCN 867
GRB 001204	3.09	20.1	GCN 898
GRB 010103	1.83	19.2	GCN 911
GRB 010119	1.13	18	GCN 919
GRB 010126	0.88	23.5	GCN 926
GRB 010214	0.83	21.3	GCN 949
GRB 010220	0.35	23.5	GCN 958
GRB 010324	1.29	22.3[a]	GCN 1024
GRB 010326A	0.50	21.5	GCN 1022
GRB 010412	0.60	20.5[a]	GCN 1039
GRB 011019	1.15	25.0	GCN 1128
GRB 011212	2.0	24.0	GCN 1324
GRB 020127	0.18	19.5	GCN 1230
GRB 020409	1.25	23.5	GCN 1362

Note. — The columns are (left to right), (1) GRB name, (2) observing time after the burst, (3) R-band limit, and (4) GCN circular reference.

[a] V-band limit

CHAPTER 10

The Host Galaxy of GRB 980703 at Radio Wavelengths — A Nuclear Starburst in a ULIRG[†]

E. Berger[a], S. R. Kulkarni[a], & D. A. Frail[b]

[a]Department of Astronomy, 105-24 California Institute of Technology, Pasadena, CA 91125, USA

[b]National Radio Astronomy Observatory, P. O. Box 0, Socorro, NM 87801

Abstract

We present radio observations of GRB 980703 at 1.43, 4.86, and 8.46 GHz for the period of 350 to 1000 days after the burst. These radio data clearly indicate that there is a persistent source at the position of GRB 980703 with a flux density of approximately 70 μJy at 1.43 GHz, and a spectral index, $\beta \approx 0.32$, where $F_\nu \propto \nu^{-\beta}$. We show that emission from the afterglow of GRB 980703 is expected to be one to two orders of magnitude fainter, and therefore cannot account for these observations. We interpret this persistent emission as coming from the host galaxy — the first example of a γ-ray burst (GRB) host detection at radio wavelengths. We find that it can be explained as a result of a star formation rate (SFR) of massive stars ($M > 5$ M$_\odot$) of ≈ 90 M$_\odot$/yr, which gives a total SFR (0.1M$_\odot$ <M< 100M$_\odot$) of ≈ 500 M$_\odot$/yr. On the basis of these data alone we cannot rule out that some fraction of the radio emission originates from an obscured active galactic nucleus. Using the correlation between the radio and far-IR (FIR) luminosities of star-forming galaxies, we find that the host of GRB 980703 is at the faint end of the class of Ultra Luminous Infrared Galaxies (ULIRGs), with $L_{FIR} \sim 10^{12}$ L$_\odot$. From the radio measurements of the offset between the burst and the host, and the size of the host, we conclude that GRB 980703 occurred near the center of the galaxy in a region of star formation. A comparison of the properties of this galaxy with radio and optical surveys at a similar redshift ($z \approx 1$) reveals that the host of GRB 980703 is an average radio-selected star-forming galaxy. This result has significant implications for the potential use of a GRB-selected galaxy sample for the study of galaxies and the IGM at high redshifts, especially using radio observations, which are insensitive to extinction by dust and provide an unbiased estimate of the SFR through the well-known radio-FIR correlation.

SECTION 10.1

Introduction

Recent studies of the properties and host galaxies of γ-ray bursts (GRBs) reveal some indirect evidence for the link between GRBs and star formation. Optical measurements of the offset distribution of GRBs

[†] A version of this chapter was published in *The Astrophysical Journal*, vol. 560, 652–658, (2001).

from their host centers appears to be consistent with the distribution of collapsars in an exponential disk, but inconsistent with the expected offset distribution of delayed binary mergers (Bloom et al. 2002a). GRB 990705 is an illustrative example of this result since HST images revealed that the burst was situated in a spiral arm, just north of an apparent star forming region (Holland et al. 2001; Bloom et al. 2002a). The absence of optical afterglows from the so-called "dark GRBs" (Djorgovski et al. 2001a) points to the association of GRBs with heavily obscured, and possibly star-forming, regions. In addition, Galama & Wijers (2001) claim high column densities toward several GRBs from X-ray observations of afterglows

Consequently, one of the pressing questions in the study of GRB host galaxies is whether they are representative of star-forming galaxies at a similar redshift. If they are, then the dust-penetrating power of GRBs and their broad-band afterglow emission offer a number of unique diagnostics of their host galaxies: the obscured star formation fraction, the ISM within the disk, the local environment of the burst itself, and global and line-of-sight extinction, to name a few.

GRB 980703, which has one of the brightest (apparent magnitude) hosts to date ($R \approx 22.6$ mag; Bloom et al. 1998a; Vreeswijk et al. 1999) offers an excellent opportunity for detailed studies. The afterglow optical and near-IR (NIR) light curves exhibited pronounced flattening about 6 days after the burst and this was attributed to an underlying bright host (Bloom et al. 1998a; Castro-Tirado et al. 1999; Vreeswijk et al. 1999). Djorgovski et al. (1998) undertook spectroscopic observations of the host and obtained a redshift of 0.966. Using three different estimators ([OII], Hα and 2800Å UV continuum) of the star formation rate (SFR), Djorgovski et al. (1998) inferred extinction-corrected SFR of 10 to 30 M$_\odot$/yr.

Here we report radio observations of GRB 980703 covering the period 350–1000 days after the burst at three frequencies: 1.43, 4.86, and 8.46 GHz. This burst has the distinction of being followed up for 1000 days; the previous record-holder was GRB 970508 (445 days; Frail et al. 2000c). The organization of the paper is as follows. We summarize the radio observations and data reduction in §10.2. In §10.3 we summarize the main observational results. In §10.4 we show that the late time radio observations require a steady component over and above the decaying afterglow observations. We argue that this component is unlikely to arise from an unobscured AGN but is instead due to star formation. In §10.5, we infer the SFR from the radio observations and compare and contrast this estimate to those derived from optical observations. Thanks to the high angular resolution and accurate astrometry of radio observations we are able to derive an accurate offset between the burst and the centroid of the host, as well as constrain the size of the radio emitting region (§10.6).

SECTION 10.2

Radio Observations

Very Large Array (VLA[1]) observations of GRB 980703 were initiated on 1998, July 4.40 UT at 4.86 GHz. All observations were obtained in the standard continuum mode with 2×50 MHz contiguous bands. We used the extra-galactic sources J2330+110, J0010+109 and J0022+061 for phase calibration and 3C48 (J0137+331) and 3C147 (J0542+498) for flux calibration. We used the Astronomical Image Processing System (AIPS) for data reduction.

Late-time observations (time after the burst, $t \gtrsim 350$ days) were co-added over a period of a few to thirty days in order to increase the overall sensitivity of each detection. This is appropriate since the expected change in the flux density from the afterglow over a few days, several hundred days after the burst, is negligible relative to the associated errors in the measurements. A log of the late-time observations and the flux density measurements are summarized in Table 10.1, and the light curves are shown in Figure 10.1. A summary of the early radio data, as well as broad-band modeling is given in Frail et al. (2003b).

---

[1] The VLA is operated by the National Radio Astronomy Observatory (NRAO), a facility of the National Science Foundation operated under cooperative agreement by Associated Universities, Inc.

Results

From Figure 10.1, we see that the late-time ($t \gtrsim 350$ days) radio light curves do not exhibit the customary power-law decay expected of afterglows but instead show flattening. Using all the measurements in Table 10.1, we find the following weighted-average flux densities for the host galaxy of GRB 980703: $F_{\nu,8.46} = 39.3 \pm 4.9$ μJy, $F_{\nu,4.86} = 42.1 \pm 8.6$ μJy, and $F_{\nu,1.43} = 68.0 \pm 6.6$ μJy. We searched for, but did not find, evidence for significant variability over the 650 day monitoring period (see Figure 10.2). From these flux densities we find that the radio spectral index is $\beta = 0.32 \pm 0.12$, where $F_\nu \propto \nu^{-\beta}$.

These radio images also allow us to determine the projected angular offset between the host galaxy and afterglow of GRB 980703. For each individual detection positions were determined from Gaussian fits, and the host-GRB offset was calculated with respect to a Very Long Baseline Array (VLBA) position that was measured to 0.0007 arcsec accuracy in each coordinate on 1998 August 2 at 8.42 GHz. These offsets are displayed in Figure 10.3 and the combined value from all observing runs is shown in the insert. We find an average offset from all measurements of -0.032 ± 0.015 arcsec in RA and 0.024 ± 0.015 arcsec in declination. The uncertainty in the position of the source is given by $\delta\theta_{\text{offset}} \approx (\theta_{synbeam}/2)/(S/N)$, where $\theta_{synbeam} \approx \lambda/B_{max}$ is the half-power synthesized beam-width, λ is the observing wavelength, B_{max} is the length of the maximum baseline, and S/N is the signal-to-noise ratio of the flux measurement.

The optical measurements of Bloom et al. (2002a) for the host of GRB 980703, give an angular offset of -0.054 ± 0.055 in RA and 0.098 ± 0.065 in declination (see insert in Figure 10.3). They conclude that GRB 980703 was not significantly offset from the center of its host galaxy, in agreement with the more accurate offset measurements in the radio.

In addition to accurate measurements of the offset, the radio observations allow us to place meaningful limits on the size of the radio-emitting region (i.e., the size of the star-forming region). We find that in our highest resolution images the source is unresolved, and therefore, based on the synthesized beam size we can derive an upper limit on the physical size of the source. For our adopted cosmological parameters (section 10.5) we find that the angular diameter distance to the source is $d_A \approx 5.4 \times 10^{27}$ cm. The full synthesized beam-width at 8.46 GHz is $\theta_{HPBW} \approx 0.27$ arcsec, which gives an upper limit of $D_{rad} = d_A\theta_{HPBW} < 2.3$ kpc on the diameter of the source.

Evidence for Host Galaxy Emission in the Radio Regime

An afterglow origin is difficult to reconcile with the properties of the late-time radio emission. From the early broad-band data it was inferred that the afterglow spectrum peaked at frequency, $\nu_m \sim 4 \times 10^{12}$ Hz at $t = 1.2$ days (Vreeswijk et al. 1999). If the explosion was spherical then we expect $\nu_m \propto t^{-3/2}$ (Sari et al. 1998; Chevalier & Li 2000). Thus the radio emission at 8.46 GHz is expected to decay for $t > 70$ days after the burst, while the emission at 1.43 GHz will decay for $t > 240$ days after the burst. If the ejecta were collimated (opening angle, θ_j), then we expect a more rapid decay, $\nu_m \propto t^{-2}$, once the bulk Lorentz factor, Γ, of the flow falls below θ_j, $\Gamma(t) \lesssim \theta_j^{-1}$ (Sari et al. 1999). In this case, we expect the radio afterglow to start decaying at even earlier times, and the flux will decay faster relative to a spherical explosion. In either case, we expect the radio afterglow to decay by at least a factor of three over the time span under consideration, $350 < t < 1000$ days.

We can clearly see from Figure 10.1 that this decay is not taking place, and the flux instead remains constant over a period of approximately 650 days. This behavior is similar to the flattening observed in the optical/NIR light curves of several GRBs (including GRB 980703), when the emission from the afterglow decays below the level of emission from the host galaxy.

Furthermore, the afterglow spectrum is expected to be a power law, $F_\nu \propto \nu^{-\beta}$, where $\beta = (p-1)/2$ and p is the power law index of the Lorentz factor distribution of the shocked electrons, $N(\gamma)d\gamma \propto \gamma^{-p}d\gamma$

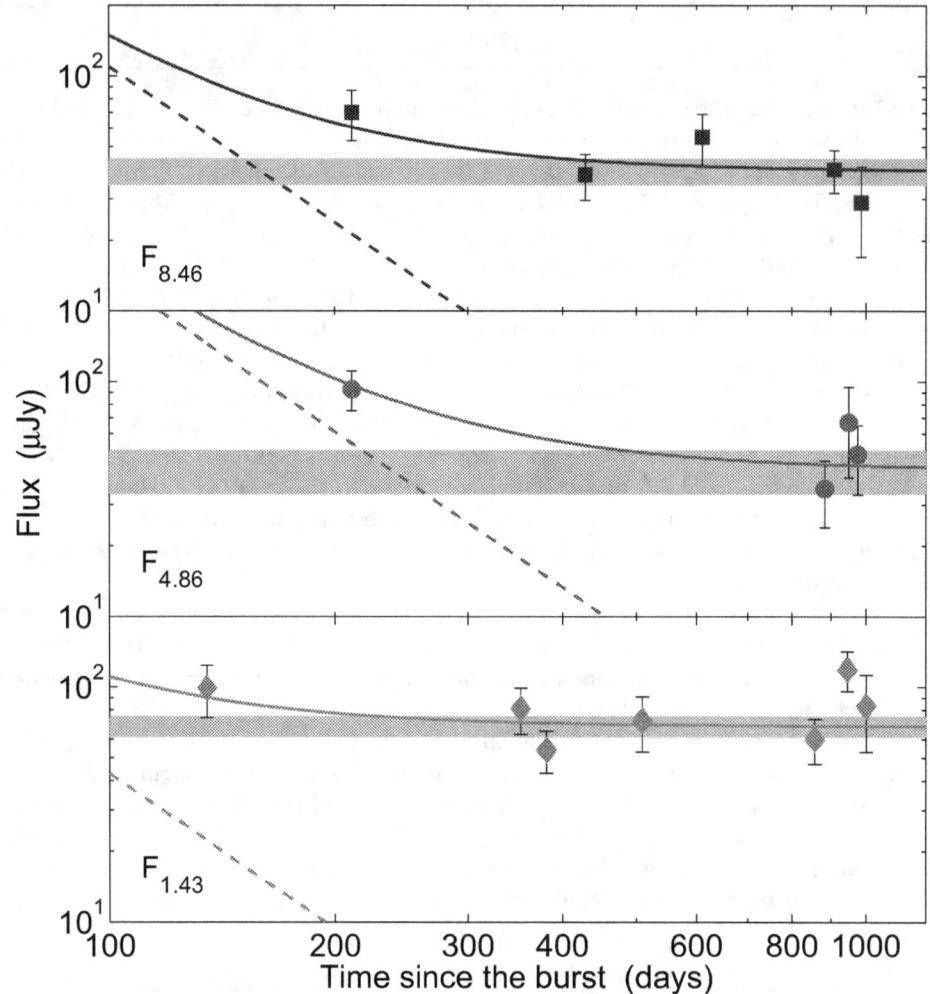

Figure 10.1: Radio light curves at 1.43, 4.86, and 8.46 GHz. The thin solid lines are the combined afterglow and host galaxy emission, the dotted lines indicate the afterglow emission, and the thick solid lines are the weighted-average fluxes of the host galaxy, with the thickness indicating the uncertainty in the flux. Only measurements at $t \gtrsim 350$ days after the burst were used to calculate the host flux. The fits are based on broad-band fitting. The data clearly indicate that there is a constant component in the observed emission, interpreted as the host galaxy.

for $\gamma > \gamma_{\min}$ (Sari et al. 1998). From the observations of many afterglows, we note that p is in the range 2.2 to 2.6 and thus we expect $\beta \sim 0.7$; for GRB 980703 the spectral index inferred from the optical data by several groups ranges from $0.6 - 1$ (Bloom et al. 1998a; Castro-Tirado et al. 1999; Vreeswijk et al. 1999). However, the observed spectral index in the range 1.43–8.46 GHz is much lower, $\beta = 0.32 \pm 0.12$. We thus conclude that there exists a steady source of emission other than the afterglow.

It is instructive to compare the characteristics of the radio emission toward GRB 980703 with those of other galaxies at a similar redshift. A radio survey of the Hubble Deep Field (HDF) and its flanking fields showed that the mean spectral index of the 8.46 GHz selected sample is $\langle \beta_{8.46} \rangle = 0.35 \pm 0.07$ (Richards 2000). In a survey of two $7' \times 7'$ fields with the VLA at 8.44 GHz Windhorst et al. (1993) found for sources with a flux density $\lesssim 100$ μJy (hereafter, μJy sources) a median spectral index,

$\beta_{\mathrm{med}} \approx 0.35 \pm 0.15$, and Fomalont et al. (1991) found $\beta_{\mathrm{med}} \approx 0.38$ for μJy sources selected at 4.9 GHz. Thus, the host galaxy of GRB 980703 appears to be a normal μJy source compared to sources selected at 4.9 or 8.5 GHz. In addition, it has been noted (Richards 2000) that the spectral index of radio sources selected at frequencies larger than 5 GHz flattens from a value of approximately 0.7 for the mJy ($F_\nu \gtrsim 1$ mJy) population to 0.3 for μJy sources.

The reason for the flattening of the spectral index may be due to the varying ratio of thermal bremsstrahlung to synchrotron emission for galaxies undergoing a burst of star formation. Supernova remnant shock acceleration of electrons results in synchrotron emission, with a characteristic spectral index of ≈ 0.8 (Condon 1992). On the other hand, thermal bremsstrahlung emission from HII regions, excited by star formation, has a much flatter spectral index, $\beta \approx 0.1$. Thus, as the direct contribution from massive stars increases the spectra are expected to flatten from a value of 0.8 to 0.1. This is exactly the effect that is observed in the aforementioned surveys.

Within the HDF and SSA13 Richards et al. (1999) identified radio sources with fluxes in the range 10–100 μJy with bright disk galaxies with $I \approx 22$ mag. The $I - K$ color for these galaxies is approximately 2.5 mag. Bloom et al. (1998a) find $I \approx 21.9$ mag and $I - K \approx 2.1$ mag for the host of GRB 980703. Thus, we see from both the radio spectrum of the source, and the optical I mag and $I - K$ color that the host galaxy of GRB 980703 has the characteristics of a typical star-forming radio galaxy selected at 8.5 GHz.

An alternate explanation for the radio emission is that it originates from an active galactic nucleus (AGN). It has been noted in surveys of the Hubble Deep Field (HDF), its flanking fields, and the Small Selected Area 13 (SSA13) that approximately 20% of the radio sources are AGN with spectral indices of about 0.3 (Richards et al. 1999; Richards 2000; Barger et al. 2000). Windhorst et al. (1993) found a similar result in their survey of two $7' \times 7'$ fields at 8.44 GHz. Thus, there is a modest probability that the emission from the host of GRB 980703 is due to an AGN.

We consider the AGN hypothesis unlikely based on the radio data and optical spectroscopy. Optical spectra of the source obtained by Djorgovski et al. (1998) show no evidence for an unobscured AGN: high-ionization lines such as Mg II $\lambda 2799$, [NeV]$\lambda 3346$, and [NeV]$\lambda 3426$ are absent, and the [OIII]$\lambda 4959$ to Hβ ratio is approximately 0.4, much lower than [OIII]/H$\beta > 1.3$ for AGN (Rola et al. 1997). Another way to discriminate between AGN and star-forming galaxies is to correlate the [OII] equivalent width (EW) with continuum color (Dressler & Gunn 1982). Kennicutt (1992) showed that AGN have redder colors for similar [OII] EW, relative to normal galaxies. Using the spectrum presented in Djorgovski et al. (1998) we evaluate the color index, $(41 - 50) \equiv 2.5\log[f_\nu(5000\text{\AA})/f_\nu(4100\text{\AA})]$ (Kennicutt 1992), and find it to be 0 ± 0.1; an AGN with the same [OII] EW would have a value $\gtrsim 0.3$ (Kennicutt 1992). Finally, Rola et al. (1997) found that for a sample of emission-line galaxies at $z \sim 0.8$, the color index between the continuum underlying the Hβ and [OII]$\lambda 3727$ lines is ≥ 0.4 for all AGN in their sample. Using the spectrum of Djorgovski et al. (1998) we find that this color index is approxmiately zero.

A second, though less persuasive argument against an AGN origin is the apparent absence of significant radio variability over the 650 day monitoring period (see Figure 10.2). The radio cores of most, but not all, low-luminosity AGN show variability exceeding the observed levels (Falcke et al. 2001).

In summary, we conclude that the radio emission seen from the host of GRB 980703 is dominated by star formation. We note, however, that the arguments presented above are not sufficient to rule out the presence of an *obscured* AGN. Determining the degree to which active star formation and/or a central engine contribute to the total emission is a common problem with centimeter and submillimeter-selected galaxies (e.g., Ivison et al. 2000). X-ray observations by *Chandra* or *XMM* may be needed to conclusively show which is the dominant power source. In the next section we assume that the bulk of the radio emisison is due to star formation and show that the host is a typical radio galaxy at $z \sim 1$ undergoing star formation.

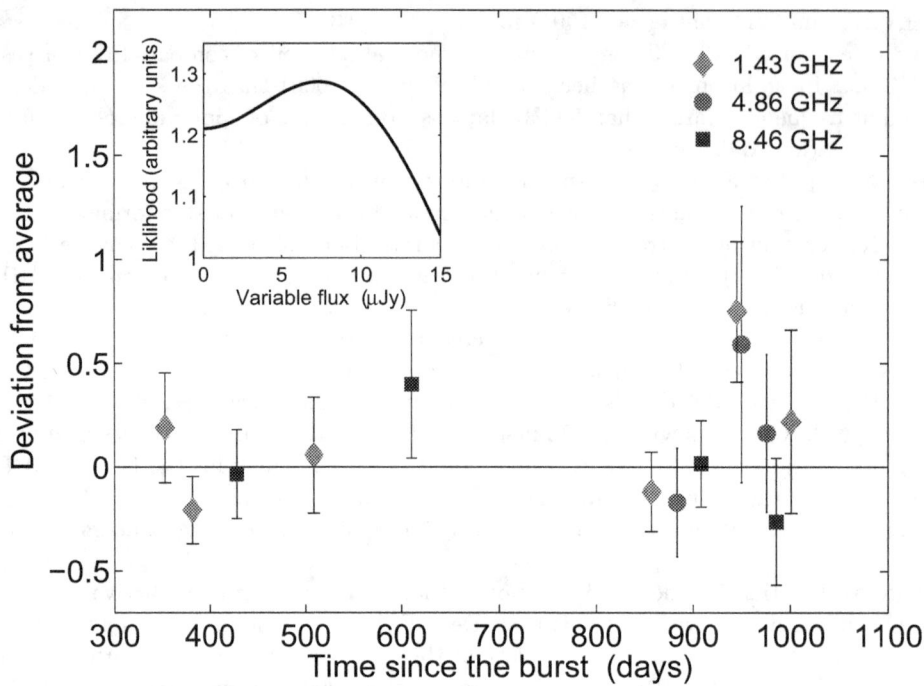

Figure 10.2: Fluctuations of individual measurements around the weighted average presented by the wide strips in figure 10.1. We note that there are no fluctuations above 1.1σ at 8.46 GHz, 0.8σ at 4.86 GHz, and 1.7σ at 1.43 GHz, indicating that the flux in each band is consistent with a constant. In fact, if we assume that the source has some variability (over that due to measurement errors), then the variable flux is less than ± 7 μJy, with a 40% probability that there are no fluctuations at all (see insert). This conclusion supports the hypothesis that the radio emission is not due to an unobscured AGN.

┌─ SECTION 10.5
│ # The Star Formation Rate in the Host Galaxy of GRB 980703
└──

Star formation is traced by optical, far-IR, submillimeter, and radio emission. In the following we will use the radio data to estimate the SFR in the host galaxy of GRB 980703, and then compare the results with the SFR derived from optical indicators, and with radio surveys at a similar redshift range in order to place the host of GRB 980703 in a larger context.

10.5.1   Star Formation Rate from the Radio Observations

Condon (1992) showed that the total radio luminosity is a combination of synchrotron and thermal emission components, both directly related to the formation rate of massive stars via a simple relationship. Moreover, since the lifetime of massive stars is of the order of 10^7 years, and the lifetime of the synchrotron emitting electrons is of the order of 10^8 years (Condon 1992), the radio emission is an excellent probe of the instantaneous SFR.

From the redshift of GRB 980703, $z = 0.966$ (Djorgovski et al. 1998), and the cosmological parameters $\Omega_0 = 0.3$, $\Lambda_0 = 0.7$ and $H_0 = 65$ km/sec/Mpc, we find that the luminosity distance to the burst is $d_L = d_A(1 + z)^2 \approx 2.1 \times 10^{28}$ cm, and the observed luminosity at each frequency is given by $L_\nu = 4\pi d_L^2 F_\nu$. The emitted luminosity is given by $L_{em,\nu'} = L_{obs,\nu}(\nu/\nu')^\beta(1 + z)^\beta$, where β is the spectral index of the radio emission, ν' is the source-frame frequency, and ν is the observing frequency. The radio spectral index derived in §10.3 is $\beta = 0.32 \pm 0.12$, and thus the emitted luminosity in each frequency is approximately 25% higher than the observed luminosity at the same frequency.

The emitted luminosity at $\nu' = 1.43$ GHz is $L_{em}(1.43) = (4.7 \pm 0.6) \times 10^{30}$ erg sec^{-1}, and we find that the SFR of massive stars in the host of GRB 980703 is

$$\text{SFR}(\text{M} > 5\text{M}_\odot) \approx \frac{L_{em}(1.43)}{5.3 \times 10^{28} \nu'^{-\beta}_{\text{GHz}} + 5.5 \times 10^{27} \nu'^{-0.1}_{\text{GHz}}} \approx 90\text{M}_\odot/\text{yr}. \tag{10.1}$$

Since both the thermal and non-thermal components are proportional only to the formation rate of high-mass stars, Equation 10.1 has to be modified by a factor which accounts for the contribution from stars in the mass range 0.1–5 M$_\odot$. For a Salpeter IMF this factor evaluates to 5.5. We use the Salpeter IMF since it is already implicitly used in equation 10.1 for the mass range 5–100 M$_\odot$. Thus, within this framework the total SFR is ≈ 500 M$_\odot$/yr.

Haarsma et al. (2000) derived the SFR for radio galaxies in the HDF, its flanking fields, SSA13, and V15 in the redshift range $0.85 - 1.15$. These fields have been observed to μJy sensitivities at cm wavelengths, and the detected radio sources have been identified with optical sources for which the redshift was determined. We use their flux and spectral index measurements along with Equation 10.1 and the correction factor to calculate the total SFR for each galaxy, and we derive a mean SFR = 657 ± 106 M$_\odot$/yr. It is clear that the host of GRB 980703 is an average star-forming radio galaxy at $z \approx 1$. This conclusion meshes well with the comparison of the radio spectral index, optical I mag, and optical $I - K$ color of the host of GRB 980703 to the same sample (see §10.4).

10.5.2 Star Formation Rate from Optical and Submillimeter Data

Djorgovski et al. (1998) used Hα and the 2800Å UV continuum to calculate a SFR of approximately 10 M$_\odot$/yr in the host of GRB 980703, after correcting for rest-frame extinction, $A_V \approx 0.3$ mag. Sokolov et al. (2001) found a similar intrinsic extinction, $A_V \approx 0.3 - 0.65$, and based on template spectral energy distributions found that the best model for the broad-band optical spectrum is given by exponentially decreasing star formation with an extinction-corrected SFR of 20 M$_\odot$/yr.

Clearly, the SFR derived from optical indicators is much lower than the value from radio measurements, even after correcting for extinction. This result is part of a general trend that has been observed in galaxies with SFR $\gtrsim 0.1$ M$_\odot$/yr (Hopkins et al. 2001). Hopkins et al. (2001) propose dust reddening dependent on SFR as the solution to this problem, and we therefore expect a much better result if we use their prescription. Extending their correlation to SFR$_{1.43} \approx 500$ M$_\odot$/yr, we find that the predicted observed SFR from Hα is approximately 70 M$_\odot$/yr. This value is still much higher than the measured SFR. In fact, in the Hopkins et al. (2001) sample the optically-derived SFR rarely exceeds 10 M$_\odot$/yr, while the radio-derived values go up to several hundred M$_\odot$/yr, indicating that the optical emission does not trace the entire star-forming volume. Thus, the SFR values for this particular galaxy are not unexpected.

The submillimeter (e.g., 350 GHz) emission from galaxies serves as another estimator of SFR, and it is related to the radio emission at 1.43 GHz via a redshift-dependent spectral index, $\beta^{350}_{1.4}$ (Carilli & Yun 1999, 2000; Dunne et al. 2000). Using the value from Carilli & Yun (2000), $\beta^{350}_{1.4} \approx 0.54 \pm 0.16$ at $z \approx 0.97$, we find $F_\nu(350) \approx 1.3^{+1.9}_{-0.8}$ mJy. Observations with the Submillimeter Common User Bolometer Array (SCUBA) camera on the James Clark Maxwell Telescope (JCMT) 12.4 days after the burst provided a 2σ upper limit of 3.2 mJy on the combined emission from the afterglow and host at 350 GHz (Smith et al. 1999), consistent with the predictions from the radio-submillimeter relation.

To conclude, we use the derived SFR to calculate the expected far-IR (FIR) emission from the host of GRB 980703. The luminosity of the FIR radiation can be derived from the empirical relation suggested by Helou et al. (1985),

$$q = -12.6 + \log(F_{FIR}/F_{1.4}) \approx 2.3 \tag{10.2}$$

which evaluates to $L_{FIR} \approx 1.5 \times 10^{12}$ L$_\odot$ for the host of GRB 980703; here F_{FIR} is the total flux in the range 40–120 μm in units of erg sec^{-1} cm^{-2}, and $F_{1.4}$ is the flux density at 1.4 GHz in units of erg sec^{-1} cm^{-2} Hz^{-1}. This value of the FIR luminosity places the host galaxy of GRB 980703 in the category of ULIRG (Sanders & Mirabel 1996). A similar claim was made for the host galaxy of GRB 970508 (Hanlon et al. 2000); however, the 41" diameter beam of the Infrared Space Observatory prevents a conclusive association of the detected 60 μm source with the host galaxy.

SECTION 10.6

Offsets and Source size

Holland et al. (2001) used a $R^{1/n}$ profile to fit the optical emission from the host and found that the best fit gives a half-light radius of 0.13 arcsec, which corresponds to an exponential disk with a scale-diameter of 0.44 arcsec. Thus, the physical size of the galaxy is $D_{opt} \approx 3.7$ kpc, 60% larger than the upper limit from our radio measurements. Holland et al. (2001) claim that the center of the galaxy is 0.2 mag bluer than the outer regions of the host. If so, star formation must be mainly taking place within the inner parts of the galaxy. Since the radio emission directly traces current star formation, we expect the radio emission to be more centrally concentrated than the optical emission. Thus, as expected, the radio size of the galaxy is smaller than the optical size.

Most likely the GRB is located within the nuclear starburst given the small offset of the GRB from the centroid of the galaxy. If so it raises the question of why the afterglow was not completely extinguished by dust. In fact, in order to reconcile the optical and radio derived SFRs we require a rest-frame extinction of $A_V \sim 4.5$ mag. Observations of the afterglow, which provide an estimate of extinction along the line-of-sight to the burst, give values of 1–2 mags from optical observations, and a somewhat higher (but highly uncertain) value from X-ray observations (Bloom et al. 1998a; Castro-Tirado et al. 1999; Vreeswijk et al. 1999). Thus, the extinction in the nuclear star-forming region is higher than the average over the whole galaxy, and the correction to the observed optical SFR is almost sufficient to reconcile it with the value of 500 M$_\odot$/yr derived from the radio.

The relatively small source size also agrees well with the classification of the host of GRB 980703 as a ULIRG exhibiting a starburst. Kennicutt (1998) showed that star formation with a rate $\gtrsim 20$ M$_\odot$/yr invariably takes place in circumnuclear regions of size 0.2–2 kpc, in the form of nuclear starburst. As a result, we expect that ULIRGs will have such size scales when traced by star formation, and the source size we measured for the host of GRB 980703 indicates that it is probably undergoing a nuclear starburst.

Finally, from the source size and offset measurement we conclude that GRB 980703 took place inside the star-forming region, providing further indirect evidence linking GRBs to massive stars

---
SECTION 10.7

Discussion and Conclusions

Late-time observations of GRB 980703 reveal a steady component, with a flux density $F_{\nu,1.43} = 68.0 \pm 6.6$ μJy and a spectral index $\beta = 0.32 \pm 0.12$. The spectral and temporal characteristics of this emission indicate that it does not arise from the afterglow itself, but rather it is the result of star formation in the host galaxy, with SFR ≈ 500 M$_\odot$/yr. This leads to the interpretation that this host galaxy is a ULIRG undergoing a starburst. In addition, the star formation is concentrated within the inner two kpc of the host, and the progenitor of GRB 980703 was positioned within this region of star formation. This conclusion lends additional support for the collapsar model.

If GRBs really come from massive stars, then they can be used to trace the star formation history of the universe (e.g., Wijers et al. 1998). In addition, GRBs and their afterglows are potentially detectable out to very high redshifts (Lamb & Reichart 2000). These propositions, taken together with the dust-penetrating power of their γ-ray emission, make GRBs a unique tool for the study of galaxies and the

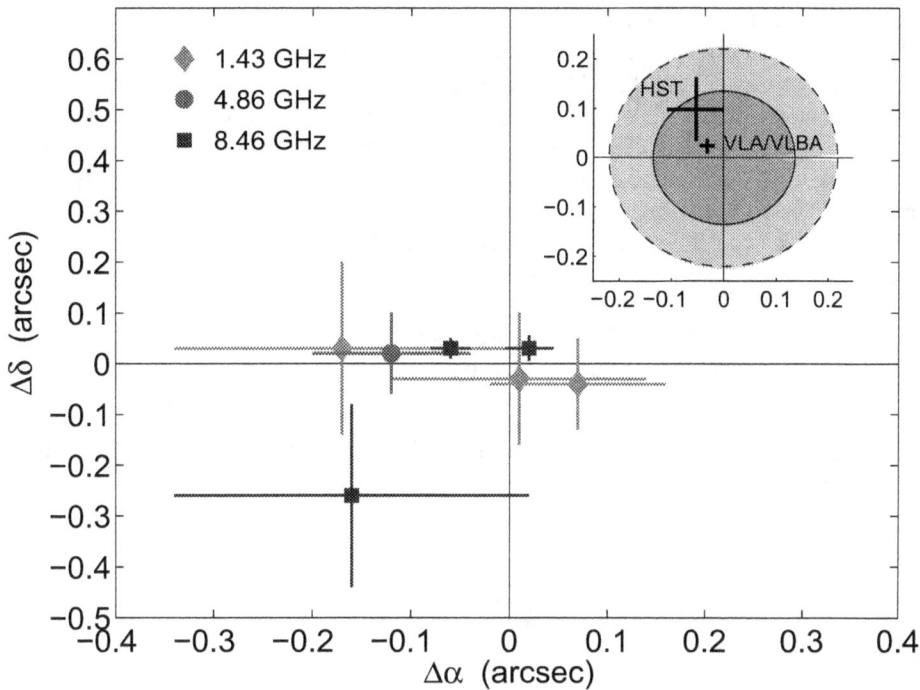

Figure 10.3: Offset measurements for all epochs in which the host galaxy emission dominates, and for which the measurements are accurate to more than 0.5 arcsec. The plot shows the offset in right ascension (α) and declination (δ) of the VLBA position of the burst (see section 10.6) relative to the host center, $(\Delta\theta_{RA}, \Delta\theta_{\delta})=(0,0)$. The most accurate measurements are at 8.46 GHz in the VLA A configuration. In this mode we achieved an rms positional error of 0.02 arcsec. The insert shows the weighted average offset in both α and δ (small cross). The larger cross is the offset measurement from Bloom et al. (2002a). The solid circle designates the projected maximum source size from the radio observations in the A configuration at 8.46 GHz, and the dashed circle is the optical size from Holland et al. (2001). Clearly the formation of massive stars is concentrated in the central region of the host. The small offset of the burst from the host center indicates that GRB 980703 occurred within the region of star formation, which points to a link between GRBs and massive stars.

IGM over a wide redshift range. In particular, radio and submillimeter/FIR observations of a GRB-selected galaxy sample will be extremely useful for the study of the obscured star formation fraction, and the properties of starbursts at high redshifts. Moreover, a comparison of the global star formation history as derived from these long-wavelength host studies, with the redshift distribution of GRBs, will provide valuable insight as to how well GRBs trace the formation rate of massive stars (e.g., Blain & Natarajan 2000); we expect that if GRBs trace only a particular channel of star formation, the two distributions will not agree.

Therefore, it is imperative to study the hosts of GRBs in the radio and submillimeter/FIR. Future observatories such as the Space Infrared Telescope Facility (SIRTF; to be launched in July 2002), the Atacama Large Millimeter Array (ALMA), the Expanded VLA (EVLA), and the Square-Kilometer Array (SKA) will allow detailed studies of these hosts. In the FIR, SIRTF will have the ability to detect sources down to several mJy, allowing the detection of galaxies with SFR comparable to that in the host of GRB 980703 out to $z \sim few$; alternatively, we will be able to detect hosts with SFR as low as a few tens of $M_\odot$/yr at $z \sim 1$. ALMA has a projected sensitivity ranging from a few μJy at 35 GHz to ~ 1 mJy at 850 GHz for a 10-minute observation, improving on the capability of an instrument such as

SCUBA by almost two orders of magnitude. Combined with an expected resolution of approximately 1 arcsec, ALMA will provide an unprecedented ability to study GRB host galaxies over a very wide redshift, SFR, and frequency range. The EVLA and SKA will greatly improve the detectability of host galaxies in the radio, and will also allow much higher angular resolution studies of compact star-forming regions. With a factor ten increase in resolution and a factor five increase in sensitivity over the current VLA, we will be able to probe scales of approximately 5 mas with the EVLA; for a galaxy at $z \approx 1$ this translates to a physical scale of 150 pc. In addition, EVLA will detect galaxies with a total SFR as low as 50 $M_\odot$/yr at $z \sim 1$. The SKA, with a similar resolution, but a much larger collecting area, will extend this capability to even lower SFR and smaller star-forming regions.

Thus, as more host galaxies are detected and studied in detail in the radio and submillimeter/FIR, we will be able to address a large number of issues pertaining not only to the bursts themselves, but also to the characteristics of galaxies at high redshifts.

We acknowledge support by NSF and NASA grants. SRK thanks Brian McBreen for useful comments on the ISO observations of GRB 970508.

Table 10.1. Late-time radio observations of GRB 980703

Epoch (UT)	Δt (days)	Config.	Time (hrs)	ν_0 (GHz)	$S \pm \sigma$ (μJy)	$\Delta\alpha$ (arcsec)	$\Delta\delta$ (arcsec)
1999 Jun. 15.36—26.29	352.65	A	6.5	1.43	81±18	−0.17 ± 0.17	0.03 ± 0.17
1999 Jul. 10.53—28.28	381.22	A	13.4	1.43	54±10	0.07 ± 0.09	−0.04 ± 0.09
1999 Aug. 19.40—Sep. 21.24	428.64	A	11.3	8.46	38±8	−0.06 ± 0.02	0.03 ± 0.02
1999 Nov. 24.06	508.88	B	1.7	1.43	72±11	0.01 ± 0.03	−0.03 ± 0.03
2000 Mar. 5.70	610.52	BnC	2.8	8.46	55±19	−0.16 ± 0.18	−0.26 ± 0.18
2000 Oct. 7.30—Nov. 19.08	847.01	A	6.5	1.43	60±13	0.01 ± 0.13	−0.03 ± 0.13
2000 Dec. 2.19—4.97	882.60	A	4.9	4.86	35±11	−0.12 ± 0.08	0.02 ± 0.08
2000 Dec. 21.15—2001 Jan. 4.98	908.38	A	7.0	8.46	40±8	0.02 ± 0.03	0.03 ± 0.03
2001 Feb. 2.00—4.93	945.29	AnB	1.9	1.43	119±23	−0.65 ± 0.30	−0.23 ± 0.30
2001 Feb. 8.08	949.90	AnB	1.2	4.86	67±28	1.22 ± 0.67	−0.38 ± 0.67
2001 Mar. 2.98—9.00	975.81	B	4.5	4.86	49±16	−1.21 ± 0.49	0.02 ± 0.49
2001 Mar. 13.97—17.98	985.80	B	4.5	8.46	29±12	0.45 ± 0.43	0.02 ± 0.43
2001 Mar. 22.96—Apr. 8.56	1001.08	B	2.6	1.43	83±30	−1.26 ± 1.76	−1.23 ± 1.76

Note. — The columns are (left to right), (1) UT date of the start of each observation or range of dates for observations which were added over several days, (2) time elapsed since the γ-ray burst, (3) array configuration, (4) total on-source observing time, (5) observing frequency, (6) peak flux density at the best fit position of the radio transient, with the error given as the root mean square noise on the image, (7) projected angular offset in RA between the host center and the position of GRB 980703, and (8) projected angular offset in declination.

CHAPTER 11

A Submillimeter and Radio Survey of Gamma-Ray Burst Host Galaxies: A Glimpse into the Future of Star Formation Studies[†]

E. Berger[a], L. L. Cowie[b], S. R. Kulkarni[a], D. A. Frail[c], H. Aussel[b] & A. J. Barger[b,d,e]

[a]Department of Astronomy, 105-24 California Institute of Technology, Pasadena, CA 91125, USA

[b]Institute for Astronomy, University of Hawaii, 2680 Woodlawn Drive, Honolulu, HI 96822

[c]National Radio Astronomy Observatory, P. O. Box 0, Socorro, NM 87801

[d]Department of Astronomy, University of Wisconsin-Madison, 475 North Charter Street, Madison, WI 53706

[e]Department of Physics and Astronomy, University of Hawaii, 2505 Correa Road, Honolulu, HI 96822

Abstract

We present the first comprehensive search for submillimeter and radio emission from the host galaxies of twenty well-localized γ-ray bursts (GRBs). With the exception of a single source, all observations were undertaken months to years after the GRB explosions to ensure negligible contamination from the afterglows. We detect the host galaxy of GRB 000418 in both the submillimeter and radio, and the host galaxy of GRB 000210 only in the submillimeter. These observations, in conjunction with the previous detections of the host galaxies of GRB 980703 and GRB 010222, indicate that about 20% of GRB host galaxies are ultra-luminous ($L > 10^{12}$ $L_\odot$) and have star formation rates of about 500 $M_\odot$ yr^{-1}. As an ensemble, the non-detected hosts have a star formation rate of about 100 $M_\odot$ yr^{-1} (5σ) based on their radio emission. This, in conjunction with an average luminosity for the entire sample that is approximately 20% fainter than the local starburst galaxy Arp 220, indicates that GRB hosts probe a more representative population of star forming galaxies than those uncovered in blank submillimeter and radio surveys. The detected and ensemble star formation rates exceed the values determined from various optical estimators by an order of magnitude, indicating significant dust obscuration. In the same vein, the ratio of bolometric dust luminosity to UV luminosity for the hosts detected in the submillimeter and radio ranges from about $\sim 30 - 500$, and follows the known trend of increasing obscuration with increasing bolometric luminosity. We also show that the GRB host sample as a whole and the submillimeter and radio detected hosts individually, have significantly bluer $R - K$ colors as

[†] A version of this chapter was published in *The Astrophysical Journal*, vol. 588, 99–112, (2003).

compared with galaxies selected in the submillimeter and radio in the same redshift range. This possibly indicates that the stellar populations in the GRB hosts are on average younger, supporting the massive stellar progenitor scenario for GRBs, but it is also possible that GRB hosts are on average less dusty. For the non-detected GRB hosts the difference in $R - K$ color may also be a manifestation of their more representative bolometric luminosities relative to the highly luminous submillimeter and radio selected galaxies. Beyond the specific results presented in this paper, the submillimeter and radio observations serve as an observational proof-of-concept in anticipation of the upcoming launch of the SWIFT GRB mission and SIRTF. These new facilities will possibly bring GRB host galaxies into the forefront of star formation studies.

SECTION 11.1

Introduction

One of the major thrusts in modern cosmology is an accurate census of star formation and star-forming galaxies in the Universe. This endeavor forms the backbone for a slew of methods (observational, analytical, and numerical) to study the process of galaxy formation and evolution over cosmic time. To date, star-forming galaxies have been selected and studied mainly in two observational windows: the rest-frame ultraviolet (UV), and rest-frame radio and far-infrared (FIR). For galaxies at high redshift these bands are shifted into the optical and radio/submillimeter, allowing observations from the ground. Still, the problem of translating the observed emission to star formation rate (SFR) involves large uncertainty. This is partly because each band traces only a minor portion of the total energy output of stars. Moreover, the optical/UV band is significantly affected by dust obscuration, thus requiring order of magnitude corrections, while the submillimeter and radio bands lack sensitivity, and therefore uncover only the most prodigiously star-forming galaxies.

The main result that has emerged from star formation surveys over the past few years is exemplified in the so-called Madau diagram. Namely, the SFR volume density, $\rho_{\text{SFR}}(z)$, rises steeply to $z \sim 1$ (Lilly et al. 1996), and seemingly peaks at $z \sim 1 - 2$. There is still some debate about the how steep the rise is, with values ranging from $(1 + z)^{1.5}$ (Wilson et al. 2002) to $(1 + z)^4$ (e.g., Madau et al. 1996). The evolution beyond $z \sim 2$ is even less clear since optical/UV observations indicate a decline (Madau et al. 1996), while recent submillimeter observations argue for a flat $\rho_{\text{SFR}}(z)$ to higher redshift, $z \sim 4$ (Barger et al. 2000). Consistency with this trend can be obtained by invoking large dust corrections in the optical/UV (Steidel et al. 1999). For general reviews of star formation surveys we refer the reader to Kennicutt (1998), Adelberger & Steidel (2000), and Blain et al. (2002).

Despite the significant progress in this field, our current understanding of star formation and its redshift evolution is still limited by the biases and shortcomings of current optical/UV, submillimeter, and radio selection techniques. In particular, despite the fact that the optical/UV band is sensitive to galaxies with modest star formation rates (down to a fraction of a $M_\odot$ yr^{-1}) at high redshift, these surveys may miss the most dusty, and vigorously star-forming galaxies. Moreover, it is not clear if the simple, locally-calibrated, prescriptions for correcting the observed *un-obscured* SFR for dust extinction (e.g., Meurer et al. 1999), hold at high redshift; even if they do, these prescriptions involve an order of magnitude correction above and beyond the inherent uncertainty in the conversion factors. Finally, the optical/UV surveys are magnitude limited, and therefore miss the faintest sources.

Submillimeter surveys have uncovered a population of highly dust-extincted galaxies, which are usually optically faint, and have star formation rates of several hundred $M_\odot$ yr^{-1} (e.g., Smail et al. 1997). However, unlike optical/UV surveys, submillimeter surveys are severely sensitivity limited, and only detect galaxies with $L_{\text{bol}} \gtrsim 10^{12}$ $L_\odot$. More importantly, current submillimeter bolometer arrays (such as SCUBA) have large beams on the sky (~ 15 arcsec) making it difficult to unambiguously identify optical counterparts (which are usually faint to begin with), and hence measure the redshifts (Smail et al. 2002); in fact, of the ~ 200 submillimeter galaxies identified to date, only a handful have a measured redshift. Finally, translating the observed submillimeter emission to a SFR requires significant assumptions about the temperature of the dust, and the dust emission spectrum (e.g., Blain et al. 2002).

Surveys at decimeter radio wavelengths also suffer from low sensitivity, but the high astrometric accuracy afforded by synthesis arrays such as the VLA allows a sub-arcsec localization of the radio-selected galaxies. As a result, it is easier to identify the optical counterparts of these sources. Recently, this approach has been used to pre-select sources for targeted submillimeter observations resulting in an increase in the submillimeter detection rate (Barger et al. 2000; Chapman et al. 2002a) and redshift determination (Chapman et al. 2003). However, this method is biased toward finding luminous (high SFR) sources since it requires an initial radio detection. An additional problem with radio, even more than with submillimeter, selection is contamination by active galactic nuclei (AGN). An examination of the X-ray properties of radio and submillimeter selected galaxies reveals that of the order of 20% can have a significant AGN component (Barger et al. 2001).

The most significant problem with current star formation studies, however, is that the link between the optical and submillimeter/radio samples is still not well understood. The Hubble Deep Field provides a clear illustration: the brightest submillimeter source does not appear to have an optical counterpart (Hughes et al. 1998), and only recently a detection has been claimed in the near-IR ($K \approx 23.5$ mag; Dunlop et al. 2002). Along the same line, submillimeter observations of the optically-selected Lyman break galaxies have resulted in very few detections (Chapman et al. 2000; Peacock et al. 2000; Chapman et al. 2002b), and even the brightest Lyman break galaxies appear to be faint in the submillimeter band (Baker et al. 2001). In addition, there is considerable diversity in the properties of optical counterparts to submillimeter sources, ranging from galaxies which are faint in both the optical and near-IR (NIR) to those which are bright in both bands (Ivison et al. 2000; Smail et al. 2002).

As a result of the unclear overlap, and the sensitivity and dust problems in the submillimeter and optical surveys, there is still strong disagreement about the fractions of global star formation in the optical and submillimeter/radio selected galaxies (e.g., Adelberger & Steidel 2000; Scott et al. 2002. It is therefore not clear if the majority of star formation takes place in ultra-luminous galaxies with very high star formation rates, or in the more abundant lower luminosity galaxies with star formation rates of a few $M_\odot$ yr^{-1}. Given the difficulty with redshift identification of submillimeter galaxies, the redshift distribution of dusty star forming galaxies remains highly uncertain.

One way to alleviate some of these problems is to study a sample of galaxies that is immune to the selection biases of current optical/UV and submillimeter/radio surveys, and which may draw a more representative sample of the underlying distribution of star-forming galaxies. The host galaxies of γ-ray bursts (GRBs) may provide one such sample.

The main advantages of the sample of GRB host galaxies are: (i) The galaxies are selected with no regard to their emission properties in any wavelength regime, (ii) the dust-penetrating power of the γ-ray emission results in a sample that is completely unbiased with respect to the global dust properties of the hosts, (iii) GRBs can be observed to very high redshifts with existing missions ($z \gtrsim 10$; Lamb & Reichart 2000), and as a result volume corrections for the star formation rates inferred from their hosts are trivial, (iv) the redshift of the galaxy can be determined via absorption spectroscopy of the optical afterglow, or X-ray spectroscopy allowing a redshift measurement of arbitrarily faint galaxies (the current record-holder is the host of GRB 990510 with $R = 28.5$ mag and $z = 1.619$; Vreeswijk et al. 2001b), and (v) since there is excellent circumstantial evidence linking GRBs to massive stars (e.g., Bloom et al. 2002a), the sample of GRB hosts may trace global star formation (Blain & Natarajan 2000).

Of course, the sample of GRB hosts is not immune from its own problems and potential biases. First, the sample is much smaller than the optical and submillimeter samples[1] (although the number of GRB hosts with a known redshift exceeds the number of submillimeter galaxies with a measured redshift). As a result, at the present it is not possible to assess the SFR density that is implied by GRB hosts, or its redshift evolution. Morever, despite the link between GRBs and massive stars, it is not clear

---

[1] Currently, the sample of GRB hosts numbers about 30 sources, and grows at a rate of about one per month. The upcoming SWIFT mission is expected to increase the rate to one per $2 - 3$ days.

whether GRB progenitors are truly representative of massive stars. In particular, a bias towards sub-solar metallicity for GRB progenitors (and hence their environments) has been discussed (MacFadyen & Woosley 1999; MacFadyen et al. 2001), but it appears that very massive stars (e.g., $M \gtrsim 35$ M$_\odot$) should produce black holes even at solar metallicity. The impact of metallicity on additional aspects of GRB formation (e.g., angular momentum, loss of hydrogen envelope) is not clear at present. Regardless of the exact details of these potential biases and problems, it is safe to conclude that GRB hosts provide a new perspective of global star formation, which is at least subject to a different set of systematic problems than the optical/UV and submillimeter approach.

To date, GRB host galaxies have mainly been studied in the optical and NIR bands. With the exception of one source (GRB 020124; Berger et al. 2002a), every GRB localized to a sub-arcsecond position has been associated with a star-forming galaxy (Bloom et al. 2002a). These galaxies range from $R \approx 22 - 29$ mag, have a median redshift, $\langle z \rangle \sim 1$, and are generally typical of star-forming galaxies at similar redshifts in terms of morphology and luminosity (Djorgovski et al. 2001b), with star formation rates from optical spectroscopy of $\sim 1 - 10$ M$_\odot$ yr^{-1}. At the same time, there are hints for higher than average ratios of [Ne III] 3869 to [O II] 3727, possibly indicating the presence of massive stars (Djorgovski et al. 2001b). Only two host galaxies have been detected so far in the radio (GRB 980703; Berger et al. 2001b) and submillimeter (GRB 010222; Frail et al. 2002).

Here we present submillimeter and radio observations of a sample of 20 GRB host galaxies, ranging in redshift from about 0.4 to 4.5 (§11.2); one of the 20 sources is detected with high significance in both the submillimeter and radio bands, and an additional source is detected in the submillimeter (§11.3). We compare the detected submillimeter and radio host galaxies to local and high-z ultra-luminous galaxies in §11.4, and derive the SFRs in §11.5. We then compare the inferred SFRs of the detected host galaxies, and the ensemble of undetected hosts, to optical estimates in §11.6. Finally, we compare the optical properties of the GRB host galaxies to those of submillimeter and radio selected star-forming galaxies (§11.7).

SECTION 11.2

Observations

11.2.1 Target Selection

At the time we conducted our survey, the sample of GRB host galaxies numbered 25, twenty of which had measured redshifts. These host galaxies were localized primarily based on optical afterglows, but also using the radio and X-ray afterglow emission. Of the 25 host galaxies we observed eight in both the radio and submillimeter, seven in the radio, and five in the submillimeter. The galaxies were drawn from the list of 25 hosts at random, constrained primarily by the availability of observing time. Thus, the sample presented here does not suffer from any obvious selection biases, with the exception of detectable afterglow emission in at least one band.

Submillimeter observations of GRB afterglows, and a small number of host galaxies have been undertaken in the past. Starting in 1997, Smith et al. (1999) and Smith et al. (2001) have searched for submillimeter emission from the afterglow of thirteen GRBs. While they did not detect any afterglow emission, these authors used their observations to place constraints on emission from eight host galaxies, with typical 1σ rms values of 1.2 mJy. Since these were target-of-opportunity observations, they were not always undertaken in favorable observing conditions.

More recently, Barnard et al. (2003) reported targeted submillimeter observations of the host galaxies of four optically-dark GRBs (i.e., GRBs lacking an optical afterglow). They focused on these particular sources since one explanation for the lack of optical emission is obscuration by dust, which presumably points to a dusty host. None of the hosts were detected, with the possible exception of GRB 000210 (see §11.3.4), leading the authors to conclude that the hosts of dark bursts are not necessarily heavily dust obscured.

Thus, the observations presented here provide the most comprehensive and bias-free search for submillimeter emission from GRB host galaxies, and the first comprehensive search for radio emission.

11.2.2 Submillimeter Observations

Observations in the submillimeter band were carried out using the Sub-millimeter Common User Bolometer Array (SCUBA; Holland et al. 1999) on the James Clerk Maxwell Telescope (JCMT[2]). We observed the positions of thirteen well-localized GRB afterglows with the long (850 μm) and short (450 μm) arrays. The observations, summarized in Table 11.1, were conducted in photometry mode with the standard nine-jiggle pattern using the central bolometer in each of the two arrays to observe the source. In the case of GRB 000301C we used an off-center bolometer in each array due to high noise levels in the central bolometer.

To account for variations in the sky brightness, we used a standard chopping of the secondary mirror between the on-source position and a position 60 arcsec away in azimuth, at a frequency of 7.8125 Hz. In addition, we used a two-position beam switch (nodding), in which the beam is moved off-source in each exposure to measure the sky. Measurements of the sky opacity (sky-dips) were taken approximately every two hours, and the focus and array noise were checked at least twice during each shift.

The pointing was checked approximately once per hour using several sources throughout each shift, and was generally found to vary by $\lesssim$ 3 arcsec (i.e., less than one quarter of the beam size). All observations were performed in band 2 and 3 weather with $\tau_{225\,\mathrm{GHz}} \approx 0.05 - 0.12$.

The data were initially reduced with the SCUBA Data Reduction Facility (SURF) following the standard reduction procedure. The off-position pointings were subtracted from the on-position pointings to account for chopping and nodding of the telescope. Noisy bolometers were removed to facilitate a more accurate sky subtraction (see below), and the data were then flat-fielded to account for the small differences in bolometer response. Extinction correction was performed using a linear interpolation between skydips taken before and after each set of on-source scans.

In addition to the sky subtraction offered by the nodding and chopping, short-term sky contributions were subtracted by using all low-noise off-source bolometers (sky bolometers). This procedure takes advantage of the fact that the sky contribution is correlated across the array. As a result, the flux in the sky bolometers can be used to assess the sky contribution to the flux in the on-source bolometer. This procedure is especially crucial when observing weak sources, since the measured flux may be dominated by the sky. We implemented the sky subtraction using SURF and our own routine using MATLAB. We found that in general the SURF sky subtraction under-estimated the sky contribution, and as a result over-estimated the source fluxes; the discrepancy in fluxes varied from about 0.1 to 0.5 mJy. Since the discrepancies were not severe, and to maintain a conservative approach we used the results of our own analysis routine. For this purpose we calculated the median value of the two (three) outer rings of bolometers in the 850 μm (450 μm) array, after removing noisy bolometers (defined as those whose standard deviation over a whole scan deviated by more than 5σ from the median standard deviation of all sky bolometers).

Following the sky subtraction, we calculated the mean and standard deviation of the mean (SDOM) for each source in a given observing shift. Noisy data were eliminated in two ways. First, the data were binned into 25 equal time bins, and the SDOM was calculated step-wise, i.e., at each step the data from an additional bin were added and the mean and SDOM were re-calculated. In an ideal situation where the data quality remains approximately constant, the SDOM should progressively decrease as more data are accumulated. However, if the quality of the data worsens (due to deteriorating weather conditions for example) the SDOM will increase. We therefore removed time bins in which the SDOM increased. After applying this procedure, we recursively eliminated individual noisy data points using a sigma cutoff level based on the number of data points (Chauvenet's criterion; Taylor 1982) until the

---

[2] The JCMT is operated by the Joint Astronomy Centre on behalf of the Particle Physics and Astronomy Research Council of the UK, the Netherlands Organization for Scientific Research, and National Research Council of Canada

mean converged on a constant value. Typically, two or three iterations were required, with only a few data points rejected each time. For all sources only a few percent of the data were rejected by the two procedures.

Finally, flux conversion factors (FCFs) were applied to the resulting voltage measurements to convert the signal to Jy. Using photometry observations of Mars and Uranus, and/or secondary calibrators (OH 231.8+4.2, IRC+10216, and CRL 618), we found the FCF to vary between $180 - 205$ Jy/V at 850 μm, consistent with the typical value of 197 ± 13. At 450 μm, the FCFs varied between $250 - 450$ Jy/V.

11.2.3 Radio Observations

Very Large Array (VLA[3] *):* We observed 12 GRB afterglow positions with the VLA from April 2001 to February 2002. All sources were observed at 8.46 GHz in the standard continuum mode with 2×50 MHz bands. In addition, GRB 000418 was observed at 1.43 and 4.86 GHz, and GRB 0010222 was observed at 4.86 GHz. In Table 11.2 we provide a summary of the observations.

In principle, since the median spectrum of faint radio sources between 1.4 and 8.5 GHz is $F_\nu \propto \nu^{-0.6}$ (Fomalont et al. 2002), the ideal VLA frequency for our observations (taking into account the sensitivity at each frequency) is 1.43 GHz. However, we chose to observe primarily at 8.46 GHz for the following reason. The majority of our observations were taken in the BnC, C, CnD, and D configurations, in which the typical synthesized beam size is $\sim 10 - 40$ arcsec at 1.43 GHz, compared to $\sim 2 - 8$ arcsec at 8.46 GHz. The large synthesized beam at 1.43 GHz, combined with the larger field of view and higher intrinsic brightness of radio sources at this frequency, would result in a significant decrease in sensitivity due to source confusion. Thus, we were forced to observe at higher frequencies, in which the reduced confusion noise more than compensates for the typical steep spectrum. We chose 8.46 GHz rather than 4.86 GHz since the combination of 20% higher sensitivity and 60% lower confusion noise, provide a more significant impact than the typical 30% decrease in intrinsic brightness. The 1.43 GHz observations of GRB 000418 were undertaken in the A configuration, where confusion does not play a limiting role.

For flux calibration we used the extragalactic sources 3C 48 (J0137+331), 3C 147 (J0542+498), and 3C 286 (J1331+305), while the phases were monitored using calibrator sources within $\sim 5°$ of the survey sources.

We used the Astronomical Image Processing System (AIPS) for data reduction and analysis. For each source we co-added all the observations prior to producing an image, to increase the final signal-to-noise.

Australia Telescope Compact Array (ATCA[4] *):* We observed the positions of four GRB afterglows during April 2002, in the 6A configuration at 1344 and 1432 MHz. Using the 6-km baseline resulted in a significant decrease in confusion noise, thus allowing observations at the most advantageous frequencies. The observations are summarized in Table 11.2.

We used J1934−638 for flux calibration, while the phase was monitored using calibrator sources within $\sim 5°$ of the survey sources. The data were reduced and analyzed using the Multichannel Image Reconstruction, Image Analysis and Display (MIRIAD) package, and AIPS.

11.2.4 Optical Data

The photometric and spectroscopic optical/NIR data used in this paper (see §11.6 and §11.7) have been collected from the literature. Host galaxy optical and NIR magnitudes are given in the Vega magnitude system. In addition, the star formation rates obtained from various optical estimators are corrected for extinction within the host galaxy when an estimate of the extinction is available (e.g., using the Balmer decrement, Djorgovski et al. 1998).

---

[3] The VLA is operated by the National Radio Astronomy Observatory, a facility of the National Science Foundation operated under cooperative agreement by Associated Universities, Inc.

[4] The Australia Telescope is funded by the Commonwealth of Australia for operations as a National Facility managed by CSIRO.

Results

The flux measurements at the position of each GRB are given in Tables 11.1 and 11.2, and are plotted in Figure 11.1. Of the 20 sources, only GRB 000418 was detected in both the radio and submillimeter with $S/N > 3$ (§11.3.1). One additional source, GRB 000210, is detected with $S/N > 3$ when combining our observations with those of Barnard et al. (2003). Two hosts have radio fluxes with $3 < S/N < 4$ (GRB 000301C and GRB 000926), but as we show below this is due in part to emission from the afterglow.

The typical 2σ thresholds are about 2 mJy, 20 μJy, and 70 μJy in the SCUBA, VLA, and ATCA observations, respectively. In Figure 11.1 we plot all sources with $S/N > 3$ as detections, and the rest as 2σ upper limits. In addition, for the sources observed with the ATCA we plot both the 1.4 GHz upper limits, and the inferred upper limits at 8.46 GHz assuming a typical radio spectrum, $F_\nu \propto \nu^{-0.6}$ (Fomalont et al. 2002).

One obvious source for the observed radio and submillimeter fluxes (other than the putative host galaxies) is emission from the afterglows. To assess the possibility that our observations are contaminated by flux from the afterglows we note that the observations have been undertaken at least a year after the GRB explosion[5] . On this timescale, the submillimeter emission from the afterglow is expected to be much lower than the detection threshold of our observations. In fact, the brightest submillimeter afterglows to date have only reached a flux of a few mJy (at 350 GHz), and typically exhibited a fading rate of $\sim t^{-1}$ after about one day following the burst (Smith et al. 1999; Berger et al. 2000; Smith et al. 2001; Frail et al. 2002; Yost et al. 2002). Thus, on the timescale of our observations, the expected submillimeter flux from the afterglows is only ~ 10 μJy, well below the detection threshold.

The radio emission from GRB afterglows is more long-lived, and hence posses a more serious problem. However, on the typical timescale of the radio observations the 8.46 GHz flux is expected to be at most a few μJy (e.g., Berger et al. 2000).

In the following sections we discuss the individual detections in the radio and submillimeter, and also provide an estimate for the radio emission from each afterglow.

11.3.1   GRB 000418

A source at the position of GRB 000418 is detected at four of the five observing frequencies with $S/N > 3$. The SCUBA source, which we designate SMM 12252+2006, has a flux density of $F_\nu(350\,\text{GHz}) \approx 3.2 \pm 0.9$ mJy, and $F_\nu(670\,\text{GHz}) \approx 41 \pm 19$ mJy. These values imply a spectral index, $\beta \approx 3.9^{+1.1}_{-1.3}$ ($F_\nu \propto \nu^\beta$), consistent with a thermal dust spectrum as expected if the emission is due to obscured star formation.

The radio source (VLA 122519.26+200611.1), is located at right ascension $\alpha(\text{J2000})=12^\text{h}25^\text{m}19.255^\text{s}$, and declination $\delta(\text{J2000})=20°06'11.10''$, with an uncertainty of 0.1 arcsec in both coordinates. This position is offset from the position of the radio afterglow of GRB 000418 (Berger et al. 2001) by $\Delta\alpha = -0.40 \pm 0.14$ arcsec and $\Delta\delta = -0.04 \pm 0.17$ arcsec (Figure 11.2). In comparison, the offset measured from Keck and *Hubble Space Telescope* images is smaller, $\Delta\alpha = -0.019 \pm 0.066$ arcsec and $\Delta\delta = 0.012 \pm 0.058$ arcsec.

VLA 122519.26+200611.1 has an observed spectral slope $\beta = -0.17 \pm 0.25$, flatter than the typical value for faint radio galaxies, $\beta \approx -0.6$ (Fomalont et al. 2002), and similar to the value measured for the host of GRB 980703 ($\beta \approx -0.32$; Berger et al. 2001b). The source appears to be slightly extended at 1.43 and 8.46 GHz, with a size of about 1 arcsec (8.8 kpc at $z = 1.119$).

The expected afterglow fluxes at 4.86 and 8.46 GHz at the time of our observations are about 5 and 10 μJy, respectively (Berger et al. 2001a). At 1.43 GHz the afterglow contribution is expected to be about 10 μJy based on the 4.86 GHz flux and the afterglow spectrum $F_\nu \propto \nu^{-0.65}$. Thus, despite the contribution from the afterglow, the radio detections of the host galaxy are still significant at better than

---

[5] The single exception is GRB 011211 for which SCUBA observations were taken $18 - 20$ days after the burst.

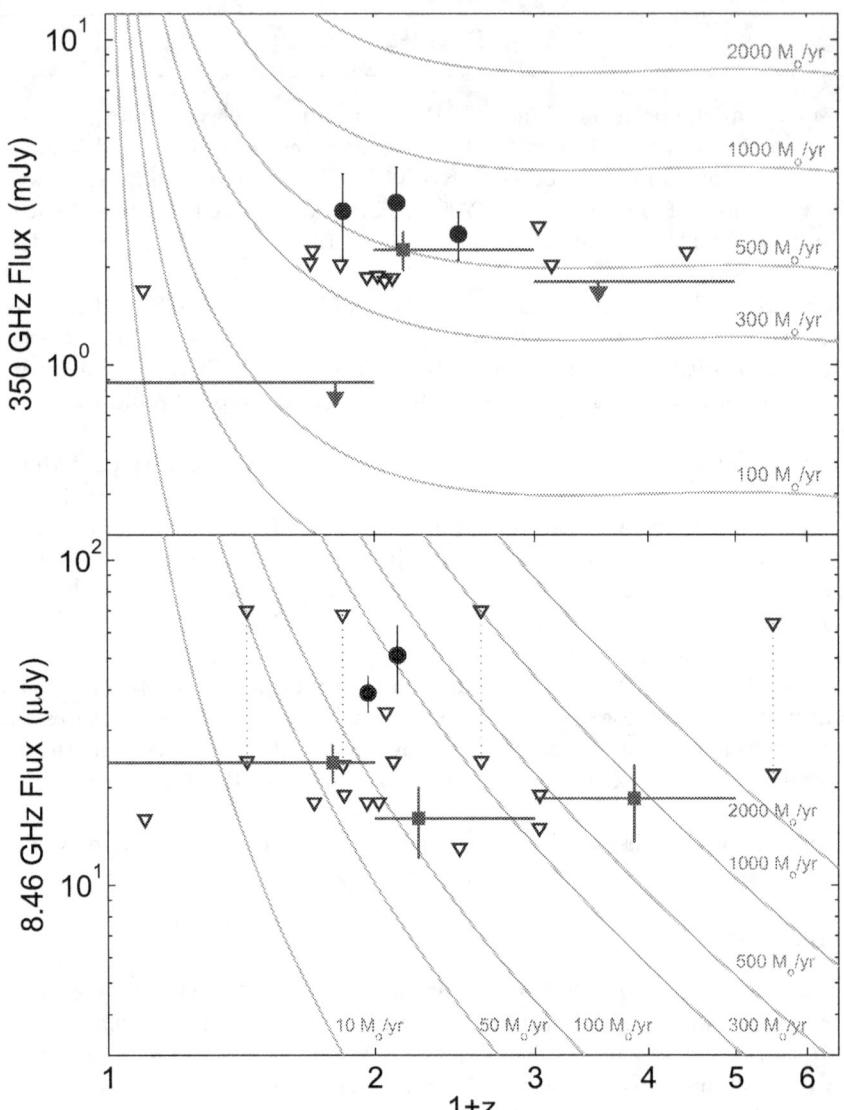

Figure 11.1: Submillimeter (top) and radio (bottom) fluxes for 20 GRB host galaxies plotted as a function of source redshift. The solid symbols are detections ($S/N > 2$ in the submillimeter and $S/N > 3$ in the radio), while the inverted triangles are 2σ upper limits. In the bottom panel, the upper limits linked by dotted lines are from the ATCA observations at 1.4 GHz (upper triangles) converted to 8.46 GHz (lower triangles) using $F_\nu \propto \nu^{-0.6}$. Also plotted are the ATCA upper limit for GRB 990712 ($z = 0.433$; Vreeswijk et al. 2001a), the VLA detection of the host of GRB 980703 (Berger et al. 2001b), and the submillimeter detection of the host of GRB 010222 Frail et al. (2002). The source at $1 + z = 1.2$ in both panels is the host of GRB 980329 which does not have a measured redshift. The points and upper limits with horizontal error bars are weighted average fluxes in the redshift bins: $0 < z < 1$, $1 < z < 2$, and $z > 2$. Finally, the thin lines are contours of constant star formation rate (using Equation 11.1 with the parameters specified in §11.5).

3σ level. Correcting for the afterglow contribution we find an actual spectral slope $\beta = -0.29 \pm 0.33$, consistent with the median $\beta \approx -0.6$ for 8.46 GHz radio sources with a similar flux (Fomalont et al.

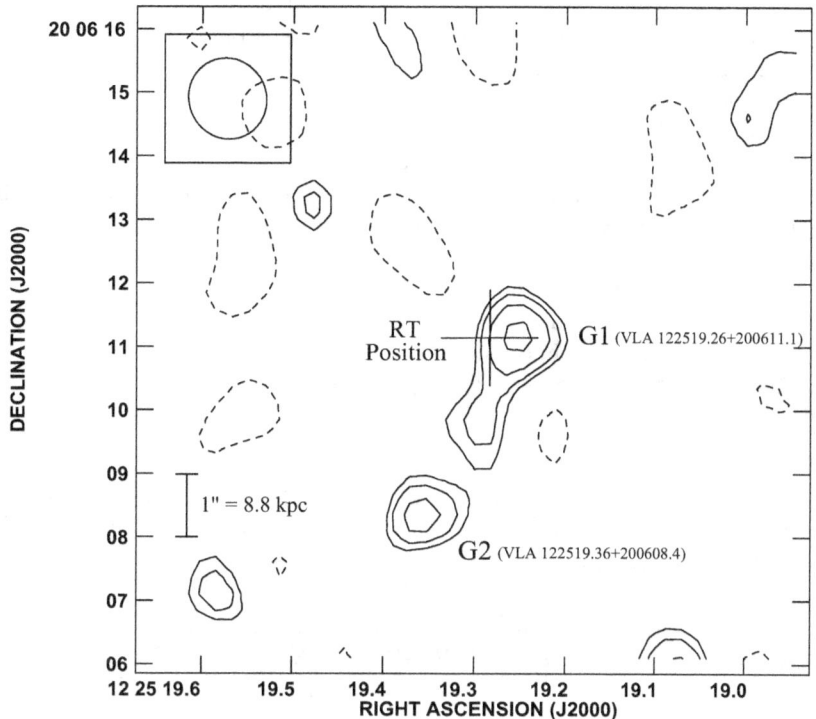

Figure 11.2: Contour plot of a 5×5 arcsec^2 field observed at 1.43 GHz and centered on the position (Berger et al. 2001b) of the radio transient associated with GRB 000418 (marked by cross). Contours are plooted at $-2^{1/2}, 2^{1/2}, 2^1, 2^{3/2}, 2^2$, and $2^{5/2}$. Source G1 is the host galaxy of GRB 000418, while source G2 is a possible companion galaxy. In addition, there appears to be a bridge of radio emission connecting galaxies G1 and G2. A comparison to the synthesized beam (upper left corner) reveals that G1 and G2 are slightly extended.

2002).

As with all SCUBA detections, source confusion arising from the large beam ($D_{\mathrm{FWHM}} \approx 14$ arcsec at 350 GHz and ≈ 6 arcsec at 670 GHz) raises the possibility that SMM 12252+2006 is not associated with the host galaxy of GRB 000418. Fortunately, the detection of the radio source, which is located 0.4 ± 0.1 arcsec away from the position of the radio afterglow of GRB 000418, indicates that SMM 12252+2006 and VLA 122519.26+200611.1 are in fact the same source — the host galaxy of GRB 000418.

Besides the positional coincidence of the VLA and SCUBA sources, we gain further confidence of the association based on the spectral index between the two bands, $\beta^{350}_{1.4}$. This spectral index is redshift dependent as a result of the different spectral slopes in the two regimes (Carilli & Yun 2000; Barger et al. 2000). We find $\beta^{350}_{1.4} \approx 0.73 \pm 0.10$, in good agreement with the Carilli & Yun (2000) value of $\beta^{350}_{1.4} = 0.59 \pm 0.16$ (for the redshift of GRB 000418, $z = 1.119$).

We also detect another source, slightly extended ($\theta \approx 1$ arcsec), approximately 1.4 arcsec East and 2.7 arcsec South of the host of GRB 000418 (designated VLA 122519.36+200608.4), with $F_\nu(1.43\,\text{GHz}) = 48 \pm 15\ \mu\text{Jy}$ and $F_\nu(8.46\,\text{GHz}) = 37 \pm 12\ \mu\text{Jy}$ (see Figure 11.2). This source appears to be linked by a bridge of radio emission (with $S/N \approx 1.5$ at both frequencies) to the host of GRB 000418. The physical separation between the two sources, assuming both are at the same redshift, $z = 1.119$, is 25 kpc. There is no obvious optical counterpart to this source in *Hubble Space Telescope* images down to about $R \sim 27.5$ mag.

Based purely on radio source counts at 8.46 GHz (Fomalont et al. 2002), the expected number of sources with $F_\nu(8.46\,\text{GHz}) \approx 37\ \mu\text{Jy}$ in a 3 arcsec radius circle is only about 2.7×10^{-4}. Thus, the coincidence of two such faint sources within 3 arcsec is highly suggestive of an interacting system, rather than chance superposition.

Interacting radio galaxies with separations of about 20 kpc, and joined by a bridge of radio continuum emission have been observed locally (Condon et al. 1993, 2002). In addition, optical surveys (e.g., Patton et al. 2002) show that a few percent of galaxies with an absolute B-band magnitude similar to that of the host of GRB 000418, have companions within about 30 kpc. The fraction of interacting systems is possibly much higher, $\sim 50\%$, in ultra-luminous systems (such as the host of GRB 000418), both locally (Sanders & Mirabel 1996) and at high redshift (e.g., Ivison et al. 2000).

We note that with a separation of only 3 arcsec, the host of GRB 000418 and the companion galaxy fall within the SCUBA beam. Thus, it is possible that SMM 12252+2006 is in fact a superposition of both radio sources. This will change the value of $\beta_{1.4}^{350}$ to about 0.46.

11.3.2 GRB 980703

The host galaxy of GRB 980703 has been detected in deep radio observations at 1.43, 4.86, and 8.46 GHz (Berger et al. 2001b). The galaxy has a flux $F_\nu(1.43\,\text{GHz}) = 68.0 \pm 6.6\ \mu\text{Jy}$, and a radio spectral slope $\beta = -0.32 \pm 0.12$. In addition, the radio emission is unresolved with a maximum angular size of 0.27 arcsec (2.3 kpc).

Based on the Carilli & Yun (2000) value of $\beta_{1.4}^{350} \approx 0.54 \pm 0.16$ (for the redshift of GRB 980703, $z = 0.966$), the expected flux at 350 GHz is $F_\nu(350\,\text{GHz}) \approx 1.3_{-0.8}^{+1.9}$ mJy. The observed (2σ) flux limit $F_\nu(350\,\text{GHz}) < 1.8$ mJy, is consistent with the expected value.

11.3.3 GRB 010222

GRB 010222 has been detected in SCUBA and IRAM observations with a persistent flux of about 3.5 mJy at 350 GHz and 1 mJy at 250 GHz (Frail et al. 2002). The persistent emission, as well as the steep spectral slope, $\beta \approx 3.8$, indicated that while the detected emission was partially due to the afterglow of GRB 010222, it was dominated by the host galaxy. In fact, accounting for the expected afterglow emission, we find that the host galaxy has a flux, $F_\nu(350\,\text{GHz}) \approx 2.5 \pm 0.4$ mJy.

The radio flux predicted from the submillimeter emission (Carilli & Yun 2000) is $F_\nu(1.43\,\text{GHz}) \approx 55_{-20}^{+80}\ \mu\text{Jy}$ (for $z = 1.477$, the redshift of GRB 010222), which corresponds to $F_\nu(4.86\,\text{GHz}) \approx 15-60\ \mu\text{Jy}$, and $F_\nu(8.46\,\text{GHz}) \approx 10-45\ \mu\text{Jy}$ (assuming $\beta = -0.6$). Therefore, our measured values, $F_\nu(4.86\,\text{GHz}) = 26 \pm 8\ \mu\text{Jy}$ and $F_\nu(8.46\,\text{GHz}) = 17 \pm 6\ \mu\text{Jy}$ are consistent with the observed submillimeter emission.

The expected afterglow fluxes at 4.86 and 8.46 GHz are 3 and 4 μJy, respectively, significantly lower than the measured values. Thus, the observed flux mainly arises from the host.

11.3.4 GRB 000210

Recently, Barnard et al. (2003) measured a flux of 3.3 ± 1.5 mJy for GRB 000210, in good agreement with our value of 2.8 ± 1.1 mJy. A weighted average (here and elsewhere we use inverse-variance weighting) of the two measurements gives $F_\nu(350\,\text{GHz}) = 3.0 \pm 0.9$ mJy, similar to the submillimeter flux from the host galaxies of GRB 000418 and GRB 010222. The radio flux at the position of GRB 000210 is

$F_\nu(8.46\,\text{GHz}) = 18 \pm 9$ μJy. Based on a redshift of 0.846 (Piro et al. 2002) and the submillimeter detection, the expected radio flux from this source (Carilli & Yun 2000) is $F_\nu(8.46\,\text{GHz}) \approx 10 - 50$ μJy, consistent with the measured value. The expected flux of the afterglow at the time of the radio observations is less than 1 μJy (Piro et al. 2002).

11.3.5   GRB 980329

Following the localization of GRB 980329, Smith et al. (1999) observed the afterglow position with SCUBA and claimed the detection of a source with a 350 GHz flux of about 5.0 ± 1.5 mJy on 1998, Apr. 5.2 UT. Subsequent observations indicated a fading trend, with a decline to 4.0 ± 1.2 mJy on Apr. 6.2, and < 1.8 mJy (2σ) on Apr. 11.2. Based on a comparison to the radio flux of the afterglow, Smith et al. (1999) concluded that the detected submillimeter flux was in excess of the emission from the afterglow itself, and therefore requires an additional component, most likely a host galaxy.

Recently, Yost et al. (2002) re-analyzed the SCUBA data and showed that the initial submillimeter flux was in fact only about 2.5 mJy, and perfectly consistent with the afterglow flux. As a result, it is not clear that an additional persistent component is required. We also re-analyzed the data from Apr. 1998 using the method described in §11.2.2. We find the following fluxes: 2.4 ± 1.0 mJy (Apr. 5), 2.4 ± 1.1 mJy (Apr. 6), 1.2 ± 0.8 mJy (Apr. 7), 1.4 ± 0.9 mJy (Apr. 8), and 1.6 ± 0.8 mJy (Apr. 11). A comparison to the results in Smith et al. (1999) reveals that, with the exception of the last epoch, they over-estimate the fluxes by about $0.5 - 2.5$ mJy.

Our observations of GRB 980329 from September and October of 2001 reveal a flux, $F_\nu(350\,\text{GHz}) = 1.8 \pm 0.8$ mJy, indicating that the flattening to a value of about 1.5 mJy in the late epochs of the Apr. 1998 observations may indicate emission from the host galaxy.

The radio observations are similarly inconclusive, with $F_\nu(8.46\,\text{GHz}) = 18 \pm 8$ μJy. We estimate that the flux of the afterglow at 8.46 GHz at the time of our observations is only $1 - 2$ μJy (Yost et al. 2002).

Since the redshift of GRB 980329 is not known, we cannot assess the expected ratio of the radio and submillimeter fluxes.

11.3.6   GRB 000926

This source is detected in the VLA observations with a flux of $F_\nu(8.46\,\text{GHz}) = 33 \pm 9$ μJy (3.7σ). The expected flux from the afterglow at the time of the observations, ≈ 420 days after the burst, is 10 μJy (Harrison et al. 2001). Thus, the observed emission exceeds the afterglow flux by 2.6σ. In the calculations below we use a host flux of 23 ± 9 μJy.

11.3.7   GRB 000301C

The VLA observations of this GRB position reveal a source with $F_\nu(8.46\,\text{GHz}) = 23 \pm 7$ μJy (3.1σ). The flux of the afterglow at the time of the observations is about 5 μJy (Berger et al. 2000). Thus, the excess emission is significant at the 2.5σ level.

The submillimeter emission predicted based on the Carilli & Yun (2000) relation is $F_\nu(350\,\text{GHz}) = 1.5^{+3.7}_{-1.1}$ mJy (for $z = 2.034$, the redshift of GRB 000301C). This value is in agreement with the measured flux of -1 ± 1.3 mJy.

SECTION 11.4

Spectral Energy Distributions

In Figure 11.3 we plot the radio-to-UV spectral energy distributions (SEDs) of the detected host galaxies of GRB 980703, GRB 000418, and GRB 010222, as well as that of Arp 220, a proto-typical local ultra-luminous IR galaxy (ULIRG; Soifer et al. 1984), and ERO J164502+4626.4 (HuR 10), a high-z analog

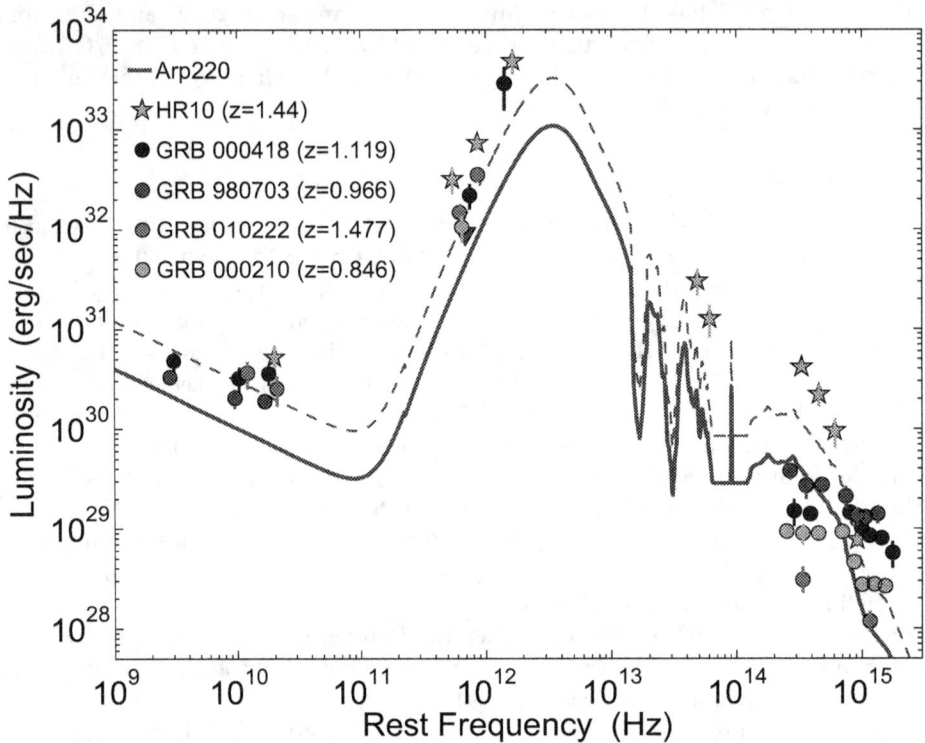

Figure 11.3: SEDs of the host galaxies of GRB 000418, GRB 980703, and GRB 010222 compared to the SED of the local starburst galaxy Arp 220, and the high-z starburst galaxy HuR 10. The luminosities are plotted at the rest frequencies to facilitate a direct comparison. The GRB host galaxies are more luminous than Arp 220, and are similar to HuR 10, indicating that their bolometric luminosities exceed 10^{12} L$_\odot$, and their star formation rates are of the order of 500 M$_\odot$ yr^{-1}. On the other hand, the spectral slopes in the optical regime are flatter than both Arp 220 and HuR 10, indicating that the GRB host galaxies are bluer than Arp 220 and HuR 10.

of Arp 220 (Hu & Ridgway 1994; Elbaz et al. 2002). The luminosities are plotted as a function of rest-frame frequencies, to facilitate a direct comparison.

The detected GRB hosts are somewhat brighter than Arp 220 ($L \approx 2 \times 10^{12}$ L$_\odot$, SFR≈ 300 M$_\odot$ yr^{-1}), and are similar in luminosity to HuR 10 ($L \approx 7 \times 10^{12}$ L$_\odot$, SFR$\sim 10^3$ M$_\odot$ yr^{-1}; Dey et al. 1999). As such, we expect the host galaxies to have star formation rates of a few $\times 100$ M$_\odot$ yr^{-1}, and bolometric luminosities in excess of 10^{12} L$_\odot$.

On the other hand, the average luminosity in the submillimeter band for all the observed hosts (detected and non-detected) is $\langle L_{\nu,s} \rangle \approx 2.1 \times 10^{31}$ erg s^{-1} Hz^{-1} at an effective rest-frame frequency of 7.9×10^{11} Hz, a factor of three less luminous than Arp 220. Similarly, in the radio band, at n effective frequency of 18 GHz, the average luminosity, $\langle L_{\nu,r} \rangle \approx 6.5 \times 10^{29}$ erg s^{-1} Hz^{-1}, is a factor of four less luminous than Arp 220.

In Figure 11.3 we use a more sophisticated approach to study the average SED of all GRB hosts in this survey, by scaling the SED of Arp 220 using a χ^2 statistic. We find that the average SED of GRB hosts in 20% fainter in both the submillimeter and radio bands than Arp 220. This clearly indicates that on average GRBs select galaxies that are less luminous than the typical submillimeter selected ULIRGs, and are therefore more representative of the general population of star-forming galaxies.

In the optical/NIR properties of the detected GRB hosts are distinctly different than those of Arp 220 and HuR 10 (as well as other local and high-z ULIRGs). In particular, from Figure 11.3 it is clear that, while the GRB host galaxies are similar to HuR 10 and Arp 220 in the radio and submillimeter bands, their optical/NIR colors (as defined for example by $R - K$) are much bluer. Moreover, while there is a dispersion of a factor of few in the radio and submillimeter bands between the GRB hosts, HuR 10, and Arp 220, the dispersion in the optical/NIR luminosity is about two orders of magnitude. This indicates that there is no simple correlation between the optical/NIR luminosities of GRB hosts (and possibly other galaxies, Adelberger & Steidel 2000) and their FIR and radio luminosities. In the following sections we expound on both points.

SECTION 11.5

Star Formation Rates

To evaluate the star formation rates that are implied by the submillimeter and radio measurements, we use the following expression for the observed flux as a function of SFR (Yun & Carilli 2002):

$$F_\nu(\nu_{\rm obs}) = \left(25 f_{\rm nth}\nu_0^{-\beta} + 0.71\nu_0^{-0.1} + 1.3 \times 10^{-6}\nu_0^3 \frac{1 - \exp[-(\nu_0/2000)^{1.35}]}{\exp(0.00083\nu_0) - 1} \right) \frac{(1 + z){\rm SFR}}{d_L^2} \ {\rm Jy.} \quad (11.1)$$

Here, $\nu_0 = (1 + z)\nu_{\rm obs}$ GHz, SFR is the star formation rate in $M_\odot$ yr^{-1}, d_L is the luminosity distance in Mpc, and $f_{\rm nth}$ is a scaling factor (of order unity) which accounts for the difference in the conversion between synchrotron flux and SFR in the Milky Way and other galaxies. The first term on the right-hand-side accounts for the fact that non-thermal synchrotron emission arising from supernova remnants is proportional to the SFR, while the second term is the contribution of free-free emission from HII regions. These two flux terms dominate in the radio regime.

The last term in Equation. 11.1 is the dust spectrum, which dominates in the submillimeter and FIR regimes. In this case, the parameters that have been chosen to characterize the spectrum are a dust temperature, $T_d = 58$ K, and a dust emissivity, $\beta = 1.35$, based on a sample of 23 IR-selected starburst galaxies with $L_{\rm FIR} > 10^{11}$ $L_\odot$ (Yun & Carilli 2002). We note that other authors (e.g., Blain et al. 2002) favor a lower dust temperature, $T_d \approx 40$ K, which would result in star formation rates that are higher by about 70%.

To calculate d_L we use the cosmological parameters $\Omega_{\rm m} = 0.3$, $\Omega_\Lambda = 0.7$, and $H_0 = 65$ km s^{-1} Mpc^{-1}. We also use the typical value $\beta \approx -0.6$ for the radio measurements (Fomalont et al. 2002). In Figure 11.1 we plot contours of constant SFR overlaid on the submillimeter and radio flux measurements. Our radio observations are sensitive to galaxies with SFR > 100 $M_\odot$ yr^{-1} at $z \sim 1$, and SFR > 1000 $M_\odot$ yr^{-1} at $z \sim 3$. The submillimeter flux, on the other hand, is relatively constant for a given SFR, independent of z. This is due to the large positive k-correction resulting from the steep thermal dust spectrum. Therefore, at the typical limit of our submillimeter observations we are sensitive to galaxies with SFR $\gtrsim 500$ $M_\odot$ yr^{-1}.

For the host galaxies that are detected with $S/N > 3$ in the submillimeter and radio, as well as those detected in the past (i.e., GRB 980703 and GRB 010222) we calculate the following star formation rates: GRB 000418 – SFR$_S = 690 \pm 200$ $M_\odot$ yr^{-1}, SFR$_R = 330 \pm 75$ $M_\odot$ yr^{-1}; GRB 000210 – SFR$_S = 560 \pm 170$ $M_\odot$ yr^{-1}; GRB 010222 – SFR$_S = 610 \pm 100$ $M_\odot$; GRB 980703 – SFR$_R = 180 \pm 25$ $M_\odot$ yr^{-1}. Here SFR$_S$ and SFR$_R$ are the SFRs derived from the submillimeter and radio fluxes, respectively. We note that the difference in the radio and submillimeter derived SFRs for GRB 000418 are an indication of the uncertainty in the dust properties and the parameter $f_{\rm nth}$.

The detections and upper limits from this survey, combined with the detections and upper limits discussed in the literature (Berger et al. 2001b; Vreeswijk et al. 2001a; Frail et al. 2002) indicate that about 20% of all GRBs explode in galaxies with star formation rates of few $\times$ 100 $M_\odot$ yr^{-1}. A similar conclusion has been reached from the shape of the 850 μm background (Barger et al. 1999). At the same time, it is clear that $\sim$ 80% of GRB host galaxies have more modest star formation rates, SFR $\lesssim$ 100

$M_\odot$ yr^{-1}.

Despite the fact that the majority of the survey sources are not detected, we can ask the question of whether the GRB host galaxies exhibit a significant submillimeter and/or radio emission *on average*. The weighted average emission from the non-detected sources ($S/N < 3$) is $\langle F_\nu(350\,\text{GHz})\rangle = 0.37 \pm 0.34$ mJy, and $\langle F_\nu(8.46\,\text{GHz})\rangle = 17.1 \pm 2.7$ μJy. This average radio flux is possibly contaminated by flux from the afterglows at the level of about 3 μJy, so we use $\langle F_\nu(8.46\,\text{GHz})\rangle \approx 14 \pm 2.7$ μJy (5.2σ). Therefore, as an ensemble, the GRB host galaxies exhibit radio emission, but no significant submillimeter emission. Using the median redshift, $z \approx 1$, for the non-detected sample, the average radio flux implies an average $\langle \text{SFR}_\text{R}\rangle \approx 100$ $M_\odot$ yr^{-1}, while the submillimeter 2σ upper limit on $\langle \text{SFR}_\text{S}\rangle$ is about 150 $M_\odot$ yr^{-1}.

The average submillimeter flux can be compared to $\langle F_\nu(350\,\text{GHz})\rangle = 0.8 \pm 0.3$ mJy for the non-detected submillimeter sources in a sample of radio pre-selected, optically faint ($I > 25$ mag) galaxies (Chapman et al. 2001), $\langle F_\nu(350\,\text{GHz})\rangle = 0.4 \pm 0.2$ mJy for Lyman break galaxies (Webb et al. 2003), or $\langle F_\nu(350\,\text{GHz})\rangle \approx 0.2$ mJy for optically-selected starbursts in the Hubble Deep Field (Peacock et al. 2000). Thus, it appears that GRB host galaxies trace a somewhat fainter population of submillimeter galaxies compared to the radio pre-selected sample, but similar to the Lyman break and HDF samples. This is not surprising given that the radio pre-selection is naturally biased in favor of luminous sources.

We can further extend this analysis by calculating the average submillimeter and radio fluxes in several redshift bins. Here we include both detections and upper limits. From the submillimeter (radio) observations we find: $\langle F_\nu\rangle = -0.2 \pm 0.4$ mJy ($\langle F_\nu\rangle = 24 \pm 3$ μJy) for $z = 0 - 1$, $\langle F_\nu\rangle = 2.3 \pm 0.3$ mJy ($\langle F_\nu\rangle = 16 \pm 4$ μJy) for $z = 1 - 2$, and $\langle F_\nu\rangle = 0.5 \pm 0.7$ mJy ($\langle F_\nu\rangle = 18 \pm 5$ μJy) for $z > 2$. These average fluxes are marked in Figure 11.1. In the submillimeter there is a clear increase in the average flux from $z < 1$ to $z \sim 1 - 2$, and a flattening or decrease beyond $z \sim 2$. In the radio, on the other hand, The average flux is about the same in all three redshift bins.

The average radio fluxes translate into the following star formation rates: for $z < 1$ the inferred average SFR is ~ 110 $M_\odot$ yr^{-1}, for $1 < z < 2$ it is ~ 200 $M_\odot$ yr^{-1}, and for $z > 2$ it is ~ 700 $M_\odot$ yr^{-1} (with $> 3\sigma$ significance in each bin). The submillimeter observations on the other hand, indicate a rise from a value of $\lesssim 160$ $M_\odot$ yr^{-1} for $z < 1$ to ~ 510 $M_\odot$ yr^{-1} for $1 < z < 2$, followed by a decline to $\lesssim 320$ $M_\odot$ yr^{-1} for $z > 2$. The difference between the two sets of SFR estimates is probably a combination of the stronger redshift dependence in the radio band, and the inherent uncertainties in the conversion factors (e.g., dust properties).

SECTION 11.6

Comparison to Optical Observations

The typical *un-obscured* star formation rates inferred from optical spectroscopy (i.e., using the UV continuum, recombination lines, and forbidden lines) are of the order of $1 - 10$ $M_\odot$ yr^{-1} (e.g., Djorgovski et al. 2001b). In particular, the host galaxy of GRB 980703 has an optical SFR of about 10 $M_\odot$ yr^{-1} (Djorgovski et al. 1998), compared to about 180 $M_\odot$ yr^{-1} from the radio observations. Similarly, the host of GRB 000418 has an optical SFR of about 55 $M_\odot$ yr^{-1} (Bloom et al. 2003a), compared to about $300 - 700$ $M_\odot$ yr^{-1} based on the radio and submillimeter detections, while the host of GRB 000210 has an optical SFR of ~ 3 $M_\odot$ yr^{-1} compared to about 550 $M_\odot$ yr^{-1} from the submillimeter observations. Finally, the average radio SFR for the non-detected sources, ~ 100 $M_\odot$ yr^{-1}, significantly exceeds the average optical SFR.

The discrepancy between the optical and radio/submillimeter star formation rates indicates that the majority of the star formation in the GRB host galaxies that are detected in the submillimeter and radio is obscured by dust. It is possible that the same is true for the sample as a whole, but this relies on the less secure average SFR in the non-detected hosts. The significant dust obscuration is not surprising given that a similar trend has been noted in high-z starburst galaxies, for which the typical dust corrections (based on the UV slope technique) are an order of magnitude (Meurer et al. 1999). In this case we find similar correction factors.

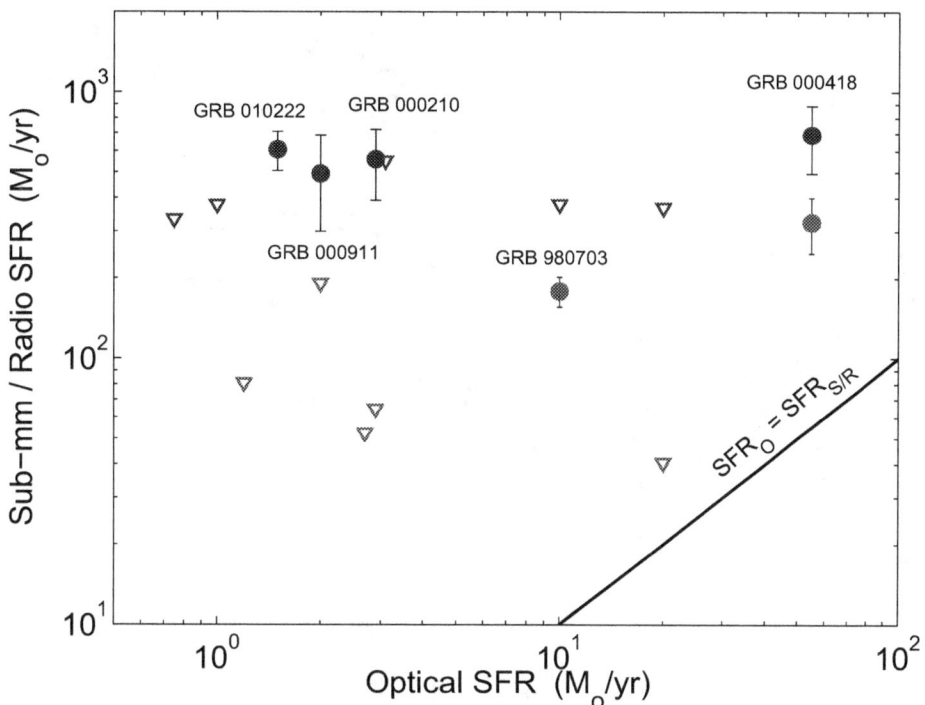

Figure 11.4: Submillimeter/radio vs. optical (i.e., from luminosity of UV continuum, recombination lines, and forbidden lines) star formation rates for several GRB host galaxies. The line in the bottom right corner designates a one-to-one correspondence between the two SFRs. Clearly, the hosts that gave appreciable submillimeter and/or radio flux have a large fraction of obscured star formation. In fact, the GRB hosts with a higher dust bolometric luminosity have a higher fraction of obscured star formation.

We can also assess the level of obscuration by comparing the UV luminosity at 1600Å, L_{1600}, to the bolometric dust luminosity, $L_{\rm bol,dust}$. The ratio of these two quantities provides a rough measure of the obscuration, while the sum provides a rough measure of the total star formation rate (Adelberger & Steidel 2000). To estimate L_{1600} we use the following host magnitudes: $B \approx 23.2$ mag (GRB 980703; Bloom et al. 1998a), $U \approx 23.5$ mag (GRB 0000210; Gorosabel et al. 2002), $R \approx 23.6$ mag (GRB 000418; Berger et al. 2001a), and $B \approx 26.7$ mag (GRB 010222; Frail et al. 2002). We extrapolate to rest-frame 1600Å using the mean value of $\langle U - R \rangle \approx 0.8$ mag found for Balmer-break galaxies, and $\langle U - R \rangle \approx 1.6$ mag found for $z \sim 1$ galaxies in the HDF that have the largest values of $L_{\rm bol,dust}$. These colors correspond to spectral slopes of -2.4 and -3.8, respectively. The resulting values of L_{1600} are: $(3.4 - 5.2) \times 10^{10}$ $L_{\odot}$ (GRB 980703), $(3.1 - 4.0) \times 10^{10}$ $L_{\odot}$ (GRB 000210), $(0.8 - 1.8) \times 10^{10}$ $L_{\odot}$ (GRB 000418), and $(0.9 - 1.0) \times 10^{10}$ $L_{\odot}$ (GRB 010222); the lower values are for $\beta = -3.8$. The mean values of L_{1600}, with the uncertainty defined as a combination of the range of reasonable spectral slopes and the intrinsic uncertainty in the host magnitudes are plotted in Figure 11.5.

We estimate $L_{\rm bol,dust}$ from the submillimeter fluxes (with the exception of GRB 980703 for which we use the radio flux) using the conversion factors given in equations 2 and 5 of Adelberger & Steidel (2000). The resulting values are: 1.3×10^{12} $L_{\odot}$ (GRB 980703), 3.3×10^{12} $L_{\odot}$ (GRB 000210), 4.4×10^{12} $L_{\odot}$ (GRB 000418), and 4.1×10^{12} $L_{\odot}$ (GRB 010222). Thus, $L_{\rm bol,dust}/L_{1600}$ evaluates to: $25 - 40$ (GRB 980703), $80 - 105$ (GRB 000210), $245 - 550$ (GRB 000418), and $410 - 455$ (GRB 010222). These results, as well as the sample of starbursts and ULIRGs at $z \sim 1$ taken from Adelberger & Steidel (2000) are plotted in Figure 11.5. We note that the GRB hosts are within the scatter of the $z \sim 1$

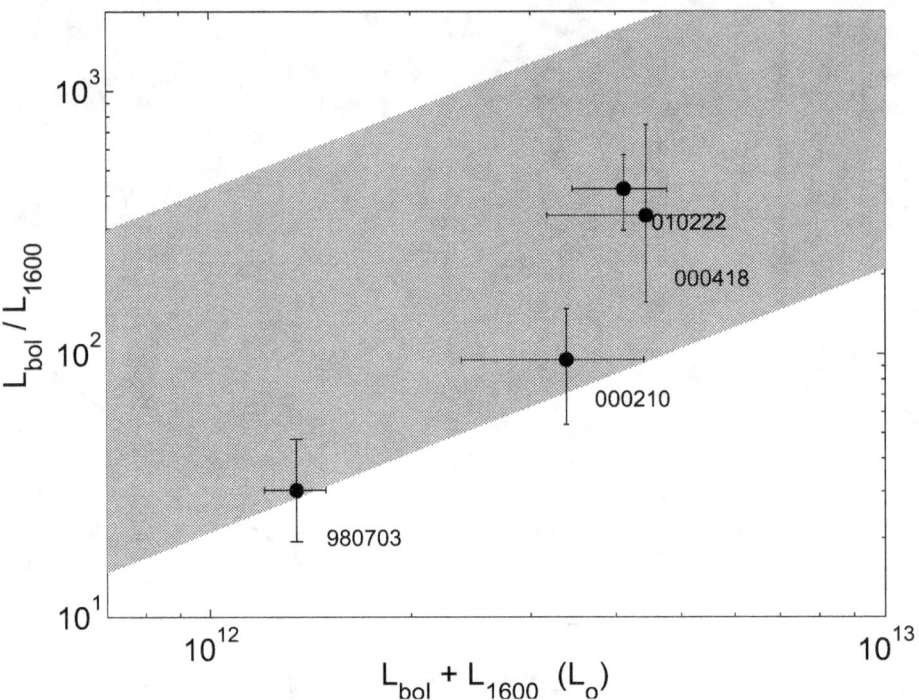

Figure 11.5: Ratio of bolometric luminosity, L_{bol} to luminosity at 1600 Å, L_{1600} plotted as a function of
the combined luminosity. The ordinate provides a measure of the amount of dust obscuration, while the
abscissa provides a measure of the total star formation rate. Black circles are the host galaxies detected
here and by Berger et al. (2001b) and Frail et al. (2002), while the shaded region is from Adelberger &
Steidel (2000) based on observations of starbursts and ULIRGs at $z \sim 1$. Clearly, there is a trend in
both samples for more dust obscuration at higher star formation rates.

sample, following the general trend of increasing value of $L_{bol,dust}/L_{1600}$ (i.e., inceasing obscuration)
with increasing $L_{bol,dust} + L_{1600}$ (i.e., inceasing SFR).

At the same time, the particular lines of sight to the GRBs within the submillimeter/radio bright
host galaxies do not appear to be heavily obscured. For example, an extinction of $A_V^{host} \sim 1$ mag has been
inferred for GRB 980703 (Frail et al. 2003b), $A_V^{host} \sim 0.4$ mag has been found for GRB 000418 (Berger
et al. 2001a), and $A_V^{host} \sim 0.1$ mag has been found for GRB 010222. The optically-dark GRB 000210
suffered more significant extinction, $A_R^{host} > 1.6$ mag. In addition, the small offset of GRB 980703
relative to its radio host galaxy (0.04 arcsec; 0.3 kpc at the redshift of the burst), combined with
the negligible extinction, indicates that while the burst probably exploded in a region of intense star
formation, it either managed to destroy a large amount of dust in its vicinity, or the dust distribution is
patchy. It is beyond the scope of this paper to evaluate the potential of dust destruction by GRBs (see
e.g., Waxman & Draine 2000), but it is clear that the GRBs that exploded in the detected submillimeter
and radio host galaxies, did not occur in the most heavily obscured star formation sites.

SECTION 11.7
Comparison of the Optical/NIR Colors of GRB hosts to Radio and Submillimeter Selected Galaxies

As we noted in §11.4, the optical/NIR colors of the detected GRB host galaxies are bluer than those of
Arp 220 ($R - K \approx 4$ mag) and HuR 10 ($I - J \approx 5.8$ mag; Dey et al. 1999). In this section we compare

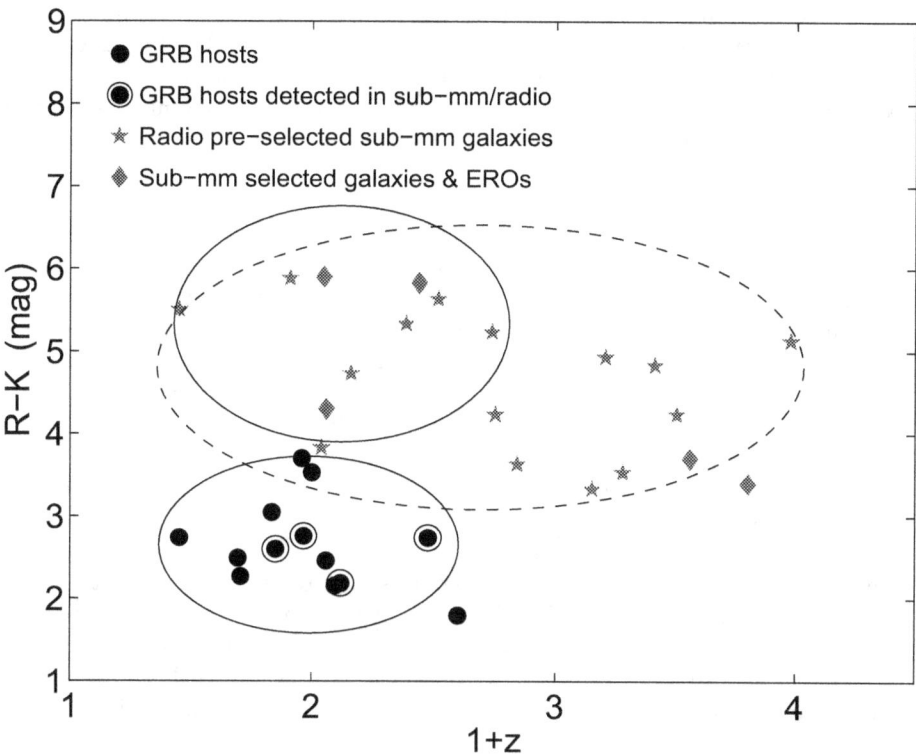

Figure 11.6: $R - K$ color as a function of redshift for GRB host galaxies, and radio pre-selected submillimeter selected (Chapman et al. 2003). The solid ellipses are centered on the mean color and redshift for each population of galaxies in the redshift range $z < 1.6$, and have widths of 2σ. The dashed ellipse is the same for the submillimeter population as a whole. Clearly, the GRB hosts are significantly bluer than the submillimeter galaxies in the same redshift range, indicating a possible preference for younger star formation episodes in GRB selected galaxies.

the $R - K$ color of GRB hosts to the $R - K$ colors of radio pre-selected submillimeter galaxies (Chapman et al. 2003) and submillimeter selected galaxies with a known optical counterpart and a redshift (Frayer et al. 1998; Ivison et al. 1998; Frayer et al. 1999).

In Figure 11.6 we plot $R - K$ color versus redshift for GRB hosts and radio pre-selected submillimeter galaxies. The optical and NIR data are collected from the literature, and are given in the Vega magnitudes. Before comparing the two populations, we note that the mean $R - K$ color and redshift for the entire GRB sample are 2.6 ± 0.6 mag and 1.0 ± 0.3, respectively, and for the hosts that are detected in the submillimeter and radio they are 2.6 ± 0.3 mag and 1.1 ± 0.3, respectively. Thus, there is no clear correlation between the optical/NIR colors of the GRB hosts and their submillimeter/radio luminosity.

For the sample of radio pre-selected and submillimeter selected galaxies the mean $R - K$ color and redshift are 4.6 ± 1.0 mag and 1.8 ± 0.7, respectively. To facilitate a more direct comparison with the GRB sample we also calculate the mean values for the same redshift range as the GRB hosts: $\langle R - K \rangle = 5.1 \pm 0.9$ mag and $\langle z \rangle = 1.1 \pm 0.3$. Clearly, the GRB host galaxies are on average significantly bluer than galaxies selected in the radio and submillimeter in the same redshift range.

Moreover, if we examine only the host galaxies that were detected in the radio and submillimeter with high significance we find $R - K$ colors of: 2.2 mag (GRB 000418), 2.8 mag (GRB 980703), 2.1 mag (GRB 010222), and 2.6 (GRB 000210). The bluest submillimeter and radio selected galaxies, on the other hand, have $R - K \approx 3.1$ mag.

The obvious difference in $R - K$ color indicates that the GRB and radio/submillimeter selections result in a somewhat different set of galaxies. The red colors of the submillimeter selected galaxies are not surprising since these sources are expected to be dust obscured. On the other hand, the mean color of the GRB hosts is bluer by about 2.5 mag (2.3σ significance) compared to submillimeter galaxies in the same redshift range, indicating a bias towards less dust obscuration. a more patchy dust distribution, or intrinsically bluer colors.

It is possible that there is a bias toward less dust obscuration in the general GRB host sample because the bursts that explode in dusty galaxies would have obscured optical afterglows, and hence no accurate localization. However, this is not a likely explanation since the GRBs which exploded in the submillimeter and radio bright hosts are not significantly dust obscured (§11.6). Moreover, it does not appear that the hosts of dark GRBs are brighter in the submillimeter as expected if the dust obscuration is global (Barnard et al. 2003). Finally, the localization of afterglows in the radio and X-rays allows the selection of host galaxies even if they are dusty. In particular, the only two GRBs in which significant obscuration of the optical afterglow has been inferred (GRB 970828: Djorgovski et al. 2001a; GRB 000210: Piro et al. 2002), have been localized thanks to accurate positions from the radio and X-ray afterglows, and have host galaxies with $R - K$ colors of 3.7 and 2.6 mag, not significantly redder than the general population of GRB hosts. Therefore, a bias against dust obscured host galaxies is not the reason for the bluer color of the sample.

An alternative explanation is that the distribution of dust in GRB hosts is different than in the radio pre-selected and submillimeter selected galaxies. This may be in terms of a spatially patchy distribution, which will allow more of the UV light to escape, or a different distribution of grain sizes (i.e., a different extinction law), possibly due to a different average metallicity. However, in both cases it is not clear why there should be a correlation between the dust properties of the galaxy and the occurence of a GRB.

Finally, it is possible that GRB host galaxies are preferentially in an earlier stage of the star formation (or starburst) process. In this case, a larger fraction of the shorter-lived massive stars would still be shining, and the overall color of the galaxy would be bluer relative to a galaxy with an older population of stars. One way to examine the age of the stellar population is to fit population synthesis models to the broad-band optical/NIR spectra of the host galaxies. This approach has recently been used by Chary et al. (2002) who find some evidence that the age of the stellar population in some GRB host galaxies (including the host of GRB 980703) is relatively young, of the order of $10 - 50$ Myr.

This result is also expected if GRBs arise from massive stars, as indicated by recent observations (e.g., Bloom et al. 2002b), since in this case GRBs would preferentially select galaxies with younger star formation episodes.

Regardless of the exact reason for the preferential selection of bluer galaxies relative to the radio pre-selected submillimeter population, two results seem clear: (i) The GRB host galaxies detected in the submillimeter and radio are likely drawn from a population that is generally missed in current submillimeter surveys, and (ii) GRB host galaxies may not be a completely bias-free sample.

The first point is particularly interesting in light of the fact that optical estimates of the SFR based on recombination and forbidden line luminosities do not identify them as particularly exceptional. Therefore, while similar galaxies are not necessarily missed in optical surveys, their star formation rates are likely under-estimated.

SECTION 11.8

Conclusions and Future Prospects

We presented the most comprehensive SCUBA, VLA, and ATCA observations of GRB host galaxies to date. The host galaxy of GRB 000418 is the only source detected with high significance in both the submillimeter and radio, while the host galaxy of GRB 000210 is detected with $S/N \approx 3.3$ in the submillimeter when we combine our observations with those of Barnard et al. (2003). When taken in conjunction with the previous detections of GRB 980703 in the radio (Berger et al. 2001b) and GRB 010222 in the submillimeter (Frail et al. 2002), these observations point to a $\sim 20\%$ detection rate in the radio/submillimeter. This detection rate confirms predictions for the number of submillimeter bright GRB hosts, with $F_\nu(350\,\mathrm{GHz}) \sim 3$ mJy, based on current models of the star formation history assuming a large fraction of obscured star formation (Ramirez-Ruiz et al. 2002).

The host galaxies detected in the submillimeter and radio have star formation rates from about 200 to 700 $\mathrm{M_\odot\ yr^{-1}}$, while statistically the non-detected sources have an *average* SFR of about 100 $\mathrm{M_\odot\ yr^{-1}}$. These star formation rates exceed the values inferred from various optical estimates by over an order of magnitude, pointing to significant dust obscuration within the GRB host galaxies detected in the submillimeter and radio, and possibly the sample as a whole.

Still, the optical afterglows of the bursts that exploded in the submillimeter/radio bright host galaxies did not suffer significant extinction, indicating that: (i) the GRBs did not explode in regions where dust obscuration is significant, or (ii) the UV and X-ray emission from the afterglow destroys a significant amount of dust in the local vicinity of the burst.

We have also shown that GRB host galaxies, even those detected in the submillimeter/radio, have bluer $R - K$ colors compared to galaxies selected in the submillimeter or radio bands in the same redshift range. This is not the result of an observational bias against dusty galaxies in the GRB host sample since the afterglows of GRBs which exploded in the radio/submillimeter bright hosts were not significantly obscured. More likely, this is the result of younger stellar populations in these galaxies, or possibly a patchy dust distribution. If the reason is younger stellar population then this provides additional circumstantial evidence in favor of massive (and hence short-lived) stars as the likely progenitors of GRBs.

A potential bias of the GRB host galaxy sample is that the popular "collapsar" model of GRBs calls for high mass, low metallicity stellar progenitors (MacFadyen & Woosley 1999). This may result in preferential selection of low metallicity (and hence less dusty) host galaxies. However, it appears that GRB progenitors can even have solar metallicity, and that a very low metallicity is unfavored by the required initial conditions for a GRB explosion. Moreover, studies of the Milky Way, local galaxies (e.g Alard 2001), and high-z galaxies (e.g., Overzier et al. 2001), indicate that there are considerable variations in metallicity within galaxies. This may be especially true if several independent episodes of star formation have occured within the galaxy. Thus, even if there is a bias towards low metallicity for GRB progenitors (and hence their immediate environments) it is not obvious that this introduces a bias in the host galaxy sample.

Nonetheless, while the observations presented in this paper clearly indicate the potential of GRB selection of high-z galaxies for the study of star formation, a much larger sample is required to complement existing optical and submillimeter surveys. This may become possible in the near future with the upcoming launch (Sep. 2003) of SWIFT. With an anticipated rapid (~ 1 minute) and accurate localization of about 150 bursts per year, the GRB-selected sample will probably increase to several hundred galaxies over the next few years. The rapid localization would most likely result in a large fraction of redshift measurements thanks to the bright optical afterglows.

In addition to the localization of a large number of GRB hosts, the study of these galaxies (as well as those in other samples) would greatly benefit from the advent of new facilities, such as SIRTF, ALMA, EVLA, and the SKA. In Figure 11.7 we again plot the rest-frame SEDs of Arp 220 and the submillimeter/radio bright GRB hosts. Overplotted on these SEDs are the 1σ sensitivities of SIRTF, ALMA, and the EVLA for 200-sec exposures at redshifts 1 and 3, as well as the sensitivities of current instruments (VLA and SCUBA).

The contributions of these new facilities to star formation studies are threefold: (i) increased sensitivity, (ii) increased resolution, and (iii) increased frequency coverage. These improvements will serve to ameliorate the main limitations of present radio, submillimeter, and IR observations (§11.1), by allowing the detection of more representative star forming galaxies at high redshift, in addition to a better constraint on the total dust bolometric luminosity and accurate localizations, which would facilitate

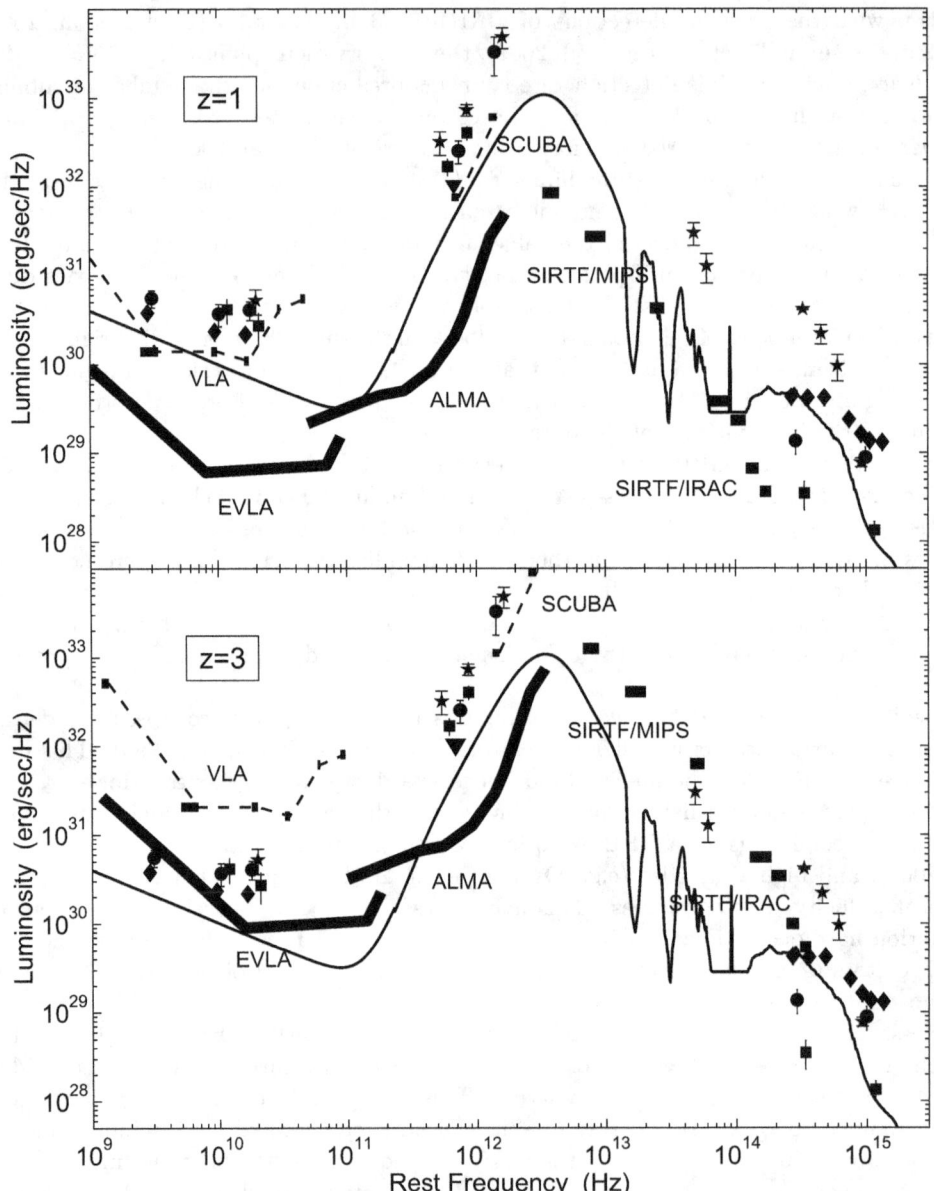

Figure 11.7: Same as Figure 11.3, overplotted with the EVLA, ALMA, and SIRTF bands at $z = 1$ and $z = 3$. The shaded regions correspond to the 1σ sensitivity in a 200 sec exposure for each instrument, while the dashed lines are the typical 1σ sensitivities for current instruments (i.e., VLA and SCUBA). Clearly, the new observatories will allow a significant increase in sensitivity, and spectral coverage over current instruments. As a result, the radio/submillimeter/IR observations will be able to probe lower luminosity (and hence more typical) star-forming galaxies.

follow-ups at optical wavelengths. In conjunction with increasingly larger samples of galaxies selected in the optical, the radio/submillimeter/IR, and by GRBs, the future of star formation studies is poised for great advances and new discoveries.

We thank A. Blain, A. Shapley, and the referee K. Adelberger, for helpful discussions and suggestions, and G. Moriarty-Schieven for help with the data reduction. We also thank S. Chapman for providing us with the optical/NIR colors and redshifts of radio pre-selected submillimeter galaxies prior to publication.

Table 11.1.   Submillimeter Observations of GRB Host Galaxies

Source	z	Obs. Date (UT)	$F_\nu(350\,\mathrm{GHz})$ (mJy)	$F_\nu(670\,\mathrm{GHz})$ (mJy)	$\langle F_\nu(350\,\mathrm{GHz})\rangle$ (mJy)
GRB 970228	0.695	Nov. 1, 2001	-1.58 ± 1.34	-21.4 ± 18.6	
		Nov. 2, 2001	0.42 ± 1.61	-10.9 ± 21.4	-0.76 ± 1.03
GRB 970508	0.835	Sep. 9, 2001	-1.70 ± 1.56	-12.2 ± 48.4	
		Sep. 10, 2001	-0.53 ± 1.60	3.2 ± 64.8	
		Sep. 12, 2001	-3.64 ± 2.43	6.0 ± 34.2	-1.57 ± 1.01
GRB 971214	3.418	Nov. 2, 2001	0.49 ± 1.11	-14.2 ± 12.6	0.49 ± 1.11
GRB 980329	—	Sep. 13, 2001	1.22 ± 1.62	8.6 ± 10.2	
		Oct. 29, 2001	2.06 ± 0.99	-27.4 ± 21.6	1.83 ± 0.84
GRB 980613	1.096	Nov. 1, 2001	2.84 ± 1.87	92.6 ± 95.9	
		Nov. 2, 2001	2.21 ± 1.77	30.3 ± 64.4	
		Dec. 7, 2001	0.93 ± 1.33	22.6 ± 17.6	1.75 ± 0.92
GRB 980703	0.966	Sep. 10, 2001	-2.40 ± 1.30	-22.6 ± 18.6	
		Sep. 12, 2001	-0.84 ± 1.33	-13.9 ± 10.7	-1.64 ± 0.93
GRB 991208	0.706	Dec. 6, 2001	-2.65 ± 1.83	9.1 ± 26.9	
		Dec. 7, 2001	-0.08 ± 1.42	26.0 ± 17.2	-1.04 ± 1.12
GRB 991216	1.020	Oct. 31, 2001	0.09 ± 1.20	-6.5 ± 21.3	
		Nov. 3, 2001	1.23 ± 1.85	-30.2 ± 31.1	
		Nov. 4, 2001	0.73 ± 2.60	25.6 ± 128.5	0.47 ± 0.94
GRB 000210	0.846	Sep. 12, 2001	3.96 ± 2.27	98.1 ± 48.2	
		Sep. 13, 2001	4.34 ± 1.63	70.1 ± 45.1	
		Sep. 14, 2001	-0.01 ± 1.87	-6.4 ± 87.1	2.97 ± 0.88
GRB 000301C	2.034	Dec. 29, 2001	1.02 ± 1.99	21.4 ± 10.7	
		Dec. 30, 2001	-2.71 ± 1.79	-18.7 ± 25.1	-1.04 ± 1.33
GRB 000418	1.119	Oct. 30, 2001	3.80 ± 2.11	9.4 ± 56.7	
		Oct. 31, 2001	3.59 ± 1.35	65.1 ± 31.4	
		Nov. 1, 2001	2.32 ± 1.46	31.9 ± 26.1	3.15 ± 0.90
GRB 000911	1.058	Sep. 13, 2001	0.56 ± 1.69	4.7 ± 22.7	
		Sep. 14, 2001	-0.37 ± 2.68	-11.1 ± 41.2	
		Oct. 31, 2001	0.95 ± 2.25	-35.0 ± 66.2	
		Nov. 3, 2001	6.73 ± 2.08	56.5 ± 52.3	
		Nov. 4, 2001	3.07 ± 1.82	-49.0 ± 51.3	2.31 ± 0.91
GRB 011211	2.140	Dec. 29, 2001	1.64 ± 1.61	8.1 ± 15.2	
		Dec. 30, 2001	-0.11 ± 1.60	-14.3 ± 42.7	
		Dec. 31, 2001	3.88 ± 2.26	17.7 ± 68.0	1.39 ± 1.01

Note. — The columns are (left to right), (1) Source name, (2) source redshift, (3) UT date for each observation, (4) flux density at 350 GHz, (5) flux density at 670 GHz, and (6) weighted-average flux density at 350 GHz.

Table 11.2.   Radio Observations of GRB Host Galaxies

Source	z	Telescope	Obs. Dates (UT)	Obs. Freq. (GHz)	F_ν (μJy)
GRB 970828	0.958	VLA	Jun. 4–7, 2001	8.46	12 ± 9
GRB 980329	—	VLA	Jul. 22 – Sep. 10, 2001	8.46	18 ± 8
GRB 980613	1.096	VLA	May 18–26, 2001	8.46	11 ± 12
GRB 981226	—	VLA	Jul. 24 – Oct. 15, 2001	8.46	21 ± 12
GRB 991208	0.706	VLA	Apr. 14 – Jul. 20, 2001	8.46	21 ± 9
GRB 991216	1.020	VLA	Jun. 8 – Jul. 13, 2001	8.46	11 ± 9
GRB 000210	0.846	VLA	Sep. 16 – Oct. 12, 2001	8.46	18 ± 9
GRB 000301C	2.034	VLA	Jun. 15 – Jul. 22, 2001	8.46	23 ± 7
GRB 000418	1.119	VLA	Jan. 14 – Feb. 27, 2002	1.43	69 ± 15
		VLA	Dec. 8, 2001 – Jan. 10, 2002	4.86	46 ± 13
		VLA	May 28 – Jun. 3, 2001	8.46	51 ± 12
GRB 000911	1.058	VLA	Mar. 21 – Apr. 2, 2001	8.46	6 ± 17
GRB 000926	2.037	VLA	Jun. 11 – Jul. 12, 2001	8.46	33 ± 9
GRB 010222	1.477	VLA	Sep. 29 – Oct. 13, 2001	4.86	19 ± 10
		VLA	Jun. 24 – Aug. 27, 2001	8.46	17 ± 6
GRB 990510	1.619	ATCA	Apr. 28, 2002	1.39	9 ± 35
GRB 990705	0.840	ATCA	Apr. 21–22, 2002	1.39	40 ± 34
GRB 000131	4.5	ATCA	Apr. 28, 2002	1.39	52 ± 32
GRB 000210	0.846	ATCA	Apr. 27–28, 2002	1.39	80 ± 52

Note. — The columns are (left to right), (1) Source name, (2) source redshift, (3) Telescope, (4) range of UT dates for each observation, (5) observing frequency, and (6) peak flux density at the position of each source.

Table 11.3.   Star Formation Rates in GRB Host Galaxies Derived from Submillimeter and Radio Observations

Source	Submm SFR ($M_\odot$ yr^{-1})	Radio SFR ($M_\odot$ yr^{-1})	Optical SFR ($M_\odot$ yr^{-1})
GRB 970228	< 335	—	1
GRB 970508	< 380	—	1
GRB 970828	—	80 ± 60	1.2
GRB 971214	120 ± 275	—	3
GRB 980329[a]	460 ± 210	615 ± 275	—
GRB 980613	380 ± 200	50 ± 140	—
GRB 980703	< 380	180 ± 25	10
GRB 981226[b]	—	150 ± 85	—
GRB 990510	—	190 ± 750	—
GRB 990705	—	190 ± 165	—
GRB 991208	< 370	70 ± 30	20
GRB 991216	< 395	80 ± 70	—
GRB 000131	—	9800 ± 6070	—
GRB 000210	560 ± 165	90 ± 45	3
GRB 000301C	< 670	640 ± 270	—
GRB 000418	690 ± 195	330 ± 75	55
GRB 000911	495 ± 195	85 ± 70	2
GRB 000926	—	820 ± 340	—
GRB 010222	610 ± 100	300 ± 115	1.5
GRB 011211	350 ± 255	—	—

Note. — The columns are (left to right), (1) Source name, (2) SFR derived from the submillimeter flux, (3) SFR derived from the radio flux, and (4) SFR derived from various optical estimators. The upper limits represent 2σ values in the case when the measured flux was negative (see Table 11.1).

CHAPTER 12

Summary and Future Directions

The Diversity of Cosmic Explosions

At this point the reader has hopefully sensed the large strides made in our understanding of gamma-ray bursts and related cosmic explosions. Over the past seven years astronomers have addressed the basic issues: the distance scale (cosmological) and the broad progenitor system (massive stars). While some of the work presented here touches on the nature of the progenitors, I have focused my attention instead on the next logical step – a detailed investigation of the energy source(s) driving cosmic explosions.

Using several observational approaches, I showed that the output of the central engine in GRBs, X-ray flashes and perhaps even GRB 980425/SN 1998bw is nearly standard, with $E_{\rm rel}$ clustered on about 10^{51} erg. This result reveals a common energy source, and hence origin, for these various explosions and sets a quantitative constraint on engine models. However, the partition of the relativistic energy varies widely, with some sources dominated by ultra-relativistic ejecta and others by mildly relativistic matter. This process presumably maps a diversity in the properties of the progenitors, for example the rotation rate of the core and the metallicity of the star. Thus, while GRBs and XRFs are exemplified by their high-energy output, the prompt energy release is a poor indicator of the total relativistic yield. Building on this understanding, and motivated by the unique properties of SN 1998bw, I also showed that the high-velocity output of type Ibc supernovae varies considerably. In fact, the local fraction of explosions that are powered by an engine is less than a few percent, suggesting that such events contribute a small fraction of the local stellar death rate.

The main question left open in these studies, is whether we are missing a significant number of events at higher redshift which would bridge the two populations. The recent discovery of GRB 031203 suggests that some diversity may exist (Soderberg et al. 2004). This burst, located at $z = 0.105$ (Prochaska et al. 2004), shares several properties with GRB 980425, most importantly an energy release of about 10^{50} erg. However, unlike GRB 980425 the energy budget is dominated by the γ-ray emission. Are we beginning to witness the extension of the standard energy result to lower energy?

While the answer is not clear at present, it is instructive to consider what selection effects are at play in the present sample. Events with a low γ-ray energy have a lower limiting volume (Figure 12.1). For example, GRB 031203 could have been detected at the BATSE sensitivity threshold only to $z \sim 0.25$, while for GRB 980425 the limiting distance is only about 100 Mpc. The fact that two such events have been detected at low redshift raises the possibility that such bursts dominate the event rate at $z \gtrsim 1$, as long as evolutionary effects are not significant. It is important to keep in mind, however, that the limit on such events of $\lesssim 3\%$ of the type Ibc supernova rate (Chapter 8) ensures that they do not exceed the rate of "classical" GRBs by more than an order of magnitude. Similarly, analyses of non-triggered BATSE bursts (Kommers et al. 2000; Stern et al. 2001) do not indicate a significant increase in the slope of the $\log N/\log S$ relation that may arise from a local homogeneous population of faint bursts.

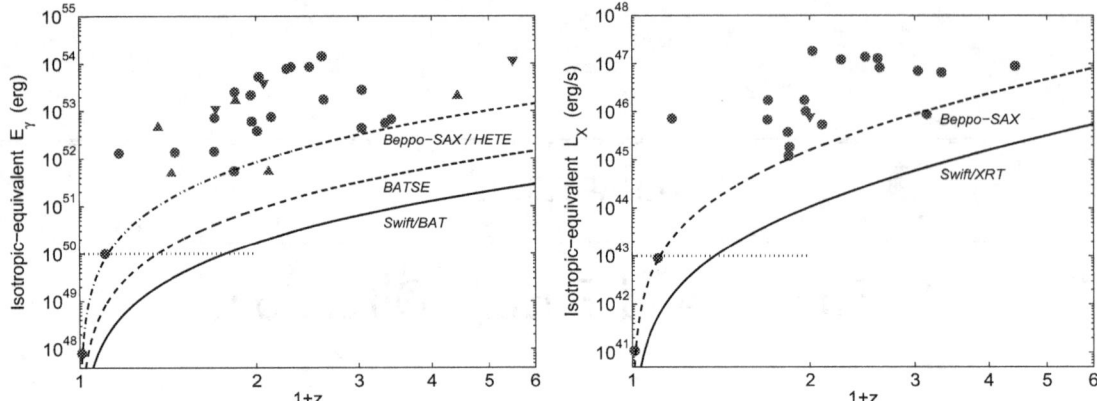

Figure 12.1: Isotropic-equivalent γ-ray energies (left) and X-ray luminosities of GRBs detected to date. The curves mark approximate detection thresholds for several missions. The increased sensitivity of *Swift* will allow the detection of events like GRB 031203 to $z \sim 1$. However, an actual redshift determination (which requires position from the afterglow) may be limited only to those events at $z \lesssim 0.4$.

In addition to the γ-ray bias, we also have to contend with a sensitivity threshold for the afterglow detections. In the X-ray band, both events could be easily detected at the distance limits determined by the γ-ray threshold. However, the bulk of such localizations in the present sample (e.g., from *Beppo-SAX*) are not sufficiently accurate for a redshift determination. In the optical and radio bands, which can provide arcsecond positions, the sensitivity threshold restricts the detection of the faint afterglows from such events to $z \lesssim 0.2$. Thus, it is possible that a sizable fraction of all GRBs lacking arcsecond positions (and hence a redshift) in fact occurred at low redshift!

A definitive answer will probably be available within the next year thanks to the launch of NASA's *Swift* satellite. This mission will overcome the selection biases detailed above in two ways. First, the γ-ray sensitivity is projected to be five times higher than that of BATSE. This will extend the limiting volume for the faintest bursts by about a factor of two. If these bursts follow $\log N/\log S \propto S^{-3/2}$ (but see e.g., Kommers et al. 2000), then the event rate will increase by as much as an order of magnitude. Perhaps more importantly, *Swift* will provide very accurate positions for the X-ray afterglows (≈ 10 arcsec) within several minutes of the burst. Thus, even in the absence of a subsequent optical and/or radio detection, a host galaxy and hence redshift could be identified. Thus, within several months of launch, *Swift* will likely allow us to determine whether the standard energy yield, as it has been determined from the current GRB sample, is in fact due to an observational bias.

SECTION 12.2
Cosmology with Gamma-Ray Bursts and Their Host Galaxies

The multi-wavelength investigation of GRB host galaxies presented in Chapters 9–11 provides an initial indication for the potential impact of GRBs on cosmological studies. The unique capabilities of *Swift* will dramatically increase the utility of GRBs as cosmological tools, both in the context of dust-obscured bursts and as lighthouses and signposts of massive star formation.

For the first time, the rapid and accurate localizations in both optical/near-IR and X-rays will remove observational bias as an impediment to the true fraction of dust obscured bursts. If the low fraction observed at the present persists, then this will likely support progenitor models that prefer low metallicity environments. For example, it has been argued in the context of collapsars (MacFadyen & Woosley 1999) that low metallicity helps keep the progenitor compact and reduces angular momentum losses from winds. However, a low metallicity also inhibits the shedding of the hydrogen envelope, suggesting that interaction with a close companion is required. Thus, the fraction of obscured bursts, while it may not provide insight into obscured star formation, will directly impact our understanding of the progenitors.

Similar insight is provided by an extension of the host galaxy work presented in Chapter 11. As mentioned in §1.6.2, GRB hosts tend to be faint in the rest-frame optical/UV. It is known however, that these bands suffer the effects of extinction and primarily provide an indication of the instantaneous star formation rate. I have therefore undertaken near-IR observations in conjunction with those published in the literature (Chary et al. 2002; Le Floc'h et al. 2003). The near-IR luminosities probe the total integrated stellar mass, since they are also sensitive to emission from old stars. The sample of GRB hosts has K-band,luminosities ranging from about -19.5 to -24 mag ($0.01 - 1$ L$_*$); see Figure 12.2. This hints at relatively low stellar masses.

A comparison of the GRB host optical and near-IR luminosities to those of other galaxy samples is illustrative. In the optical bands, GRB hosts generally have the same magnitude as a function of redshift as galaxies in the Hubble Deep Field (HDF; Cohen et al. 2000) or the Lyman break galaxies (LBGs; Shapley et al. 2003; Steidel et al. 2003, 2004). In the near-IR bands, on the other hand, GRB hosts are significantly fainter than most LBGs and all of the submillimeter-selected galaxies. However, they do have a similar magnitude distribution as a function of redshift compared to K-selected galaxies in the Subaru Deep Field (Kashikawa et al. 2003). Unfortunately, the latter only have photometric redshifts with $\delta z \gtrsim 0.5$ at faint fluxes. Clearly, the reason for the blue colors of GRB hosts is a low K-band luminosity rather than dust obscuration.

The absolute rest-frame luminosities are shown in Figure 12.2 in comparison to LBGs and submillimeter galaxies. The separation of the three samples is clear, with GRB hosts being significantly fainter in the near-IR and somewhat fainter in the optical. The rest-frame near-IR luminosities are generally thought to be related to the total mass of the galaxy since they trace light from old stellar populations, while the rest-frame optical is more sensitive to current star formation. If this is the case, then GRB hosts are likely less massive than LBGs and submillimeter galaxies, probably because they are in the initial phase of the star formation process. This will also result in somewhat lower optical luminosities, in agreement with the observed distribution.

We are therefore led to the following picture of GRB hosts. These galaxies are generally young, undergoing a first episode of starburst activity, and as a result tend to be less massive and possibly metal poor. A fraction of about 10%, however, have enough dust and a high star formation rate to produce a signal in the submillimeter band, but those are still less massive than the typical systems selected in the submillimeter. Thus, GRB selection appears to favor young starburst galaxies. This supports the inferences made based on the submillimeter and radio emission from GRB hosts. Therefore, one of the main scientific questions that GRB hosts can uniquely address are the processes that initiate the starburst process. Since these galaxies are detected at redshifts ranging at least from 0.1 to 4.5, the redshift evolution of this process may also be elucidated.

As a final note on the properties of GRB host galaxies I return to a point made earlier in this thesis. The current limit on spectroscopic redshift determination is $R \sim 25.5$ mag. Photometric redshifts extend this limit significantly, but there is no way to assess how accurate they are at low flux levels. On the other hand, GRB host galaxies of arbitrary brightness can have spectroscopic redshift measurements from absorption of the afterglow light as it traverses the galaxy. This technique also provides insight into the metallicity and dynamical state of the ISM of the host. As a result, we now have redshifts for six galaxies with $R > 26$ mag, which extend the luminosity function of high redshift galaxies an order of magnitude fainter (Figure 12.2). The volumetric corrections are difficult to assess for the GRB hosts, especially since selection effects (including γ-ray sensitivity threshold) have not been fully quantified. As a result, it is difficult to make a direct comparison with known luminosity functions. However, it is clear that there is a sizable population of galaxies, about a third of all GRB hosts, with luminosities well below $0.1L_*$.

The rapid localizations and rate of about 100 bursts per year from *Swift* will elevate the afterglows

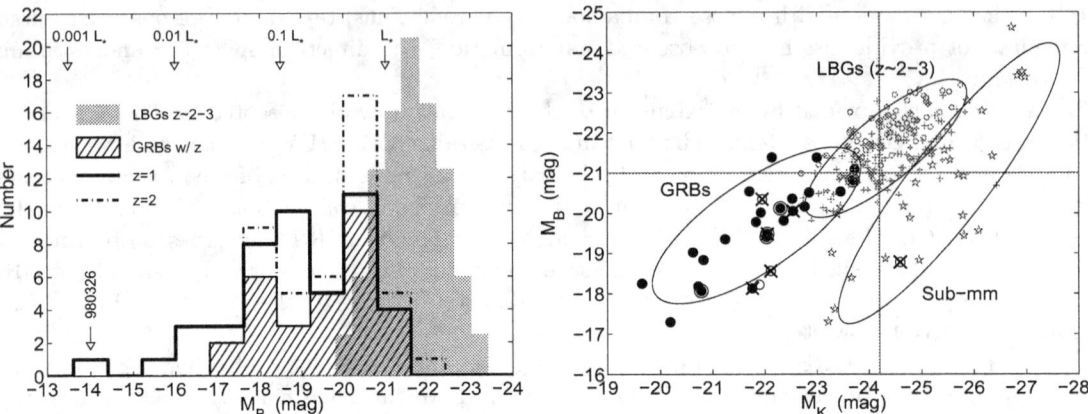

Figure 12.2: *Left:* Histogram of absolute rest-frame B-band luminosities for GRB host galaxies and Lyman break galaxies (LBGs). Gamma-ray burst hosts are typically fainter than $0.5L_*$, and they extend at least an order of magnitude fainter than LBGs. Of particular interest in the host of GRB 980326 for which Bloom et al. (1999) claim $z \lesssim 1$ based on strong evidence for an associated supernova. In this case, the host is nearly a factor of 1,000 less luminous than an L_* galaxy. *Right:* Absolute rest-frame B- versus K-band luminosities. The vertical and horizontal lines mark L_* galaxies. Submillimeter galaxies have the highest integrated stellar masses (i.e., K-band luminosities), while LBGs typically have the highest instantaneous star formation rates (i.e., B-band luminosities). Gamma-ray burst hosts are clearly separated and sub-L_* in both bands.

and host galaxies of GRBs to the forefront of IGM, ISM and star formation studies. At the present, the IGM is primarily studied using absorption spectroscopy of bright background quasars, which have revealed a filamentary structure with a wide range of column densities, and metal enrichment out to a redshift, $z \sim 4$ (Rauch 1998; Storrie-Lombardi & Wolfe 2000). Moreover, the highest redshift quasars (Becker et al. 2001), along with results from the *Wilkinson Microwave Anisotropy Probe* (Spergel et al. 2003), indicate that the Universe was re-ionized at $z \sim 7$ to 15. Unfortunately, IGM studies are limited by the ionizing effect of quasars on their local IGM (the "proximity effect"), the possibility that quasars can only probe the IGM to $z \sim 7$ (super-massive black holes possibly take several hundred million years to assemble), and the dust extinction associated with the highest density regions.

In the same vein, studies of the ISM in high redshift galaxies and its interplay with the IGM, which probes the role of feedback processes (e.g., galactic super-winds), are currently limited to the bright end of the galaxy luminosity function (e.g., LBGs with $L \gtrsim L_*$; (Adelberger et al. 2003)). Even these galaxies are typically not bright enough to elucidate the physical extent, velocity dispersion and covering fraction of super-winds, or their relation to metallicity, star formation and galactic mass.

Gamma-ray burst optical/near-IR afterglows are a unique and powerful tool in this context. The short-lived bursts (durations $\lesssim 100$ s) do not suffer from a proximity effect on scales larger than ~ 10 pc. Thus, the Lyα damping wing and metal systems near the host galaxy can be measured directly. Equally important, GRBs are at least as bright as quasars within the first few hours (Figure 1.6). This, along with explosion sites within the disks of high redshift galaxies (Bloom et al. 2002a), ensure that they can probe the ISM of *arbitrarily faint galaxies* over a wide range of redshifts. The latter is an important point since other studies relying on galaxy spectroscopy are limited to $R \sim 25$ mag.

Gamma-ray bursts can also trace denser regions than quasars since the selection trigger (γ-rays) is impervious to dust and the afterglows reside within galactic disks. Preliminary studies reveal that damped Lyα systems associated with GRB host galaxies have the highest column densities observed to date (e.g., (Savaglio et al. 2003)), specifically for this reason (Figure 12.3). Along with quasars, which tend to probe the extended halos of intervening galaxies, GRB afterglows could provide a complete

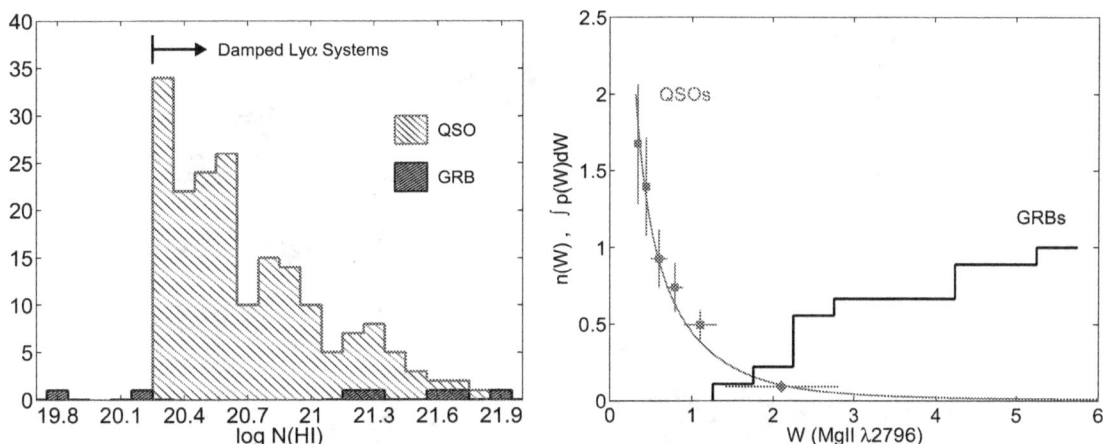

Figure 12.3: *Left:* Column densities of damped Lyα systems from QSO and GRB absorption spectra; data are from Curran et al. (2002) and Vreeswijk & et al. (2004). *Right:* Distribution of MgII equivalent widths from QSO (Steidel & Sargent 1992) and GRB afterglow spectroscopy. Clearly, GRBs trace significantly denser environments compared to quasars since they more easily probe the inner disks of high redshift systems (i.e., their host galaxies).

picture of the structure, metal distribution, and column density distribution as a function of both distance from the center of the galaxy and redshift.

Finally, the most exciting prospect is that GRBs may probe the Universe to a higher redshift than quasars since it is now thought that the first generation of massive stars may have formed beyond $z \sim 10$ (e.g., Barkana & Loeb 2001). Such bursts will probe the epoch of re-ionization with much greater precision than current studies and will provide otherwise inaccessible information about the structure of the IGM at $z \gtrsim 7$. In particular, an near-IR spectrum could provide a measurement of the Lyα optical depth, through the shape of the Lyα damping wing, and simultaneously trace the metallicity, through a measurement of the optical depth due to oxygen (e.g., using OI 1302Å; Oh 2002). The expected number of very high redshift bursts is a matter of speculation, both because the epoch of formation of massive stars is unknown and because it is not clear if these putative stars will even give rise to GRBs.

The ability of GRB afterglows to trace the ISM of their host galaxies is particularly powerful in the context of host galaxy studies. One of the main avenues of research at present is the interplay between galaxies and the IGM, especially the process of metal enrichment and the initial formation of stars. Current studies appear to favor a scenario in which galactic winds, presumably driven by supernovae, enrich the IGM (Adelberger et al. 2003). However, the faintness of the galaxies typically prevents a detailed physical understanding of how these winds arise and what influences their strength and duration. We can overcome this problem with a combination of GRB host galaxy spectroscopy and multi-wavelength imaging, providing information on star formation, and afterglow absorption spectroscopy, providing estimates of the metallicity within the galaxy, in the interface between the ISM and IGM, and in the IGM itself.

┌─ SECTION 12.3 ──
│
│   Conclusions
│
└──

The study of gamma-ray bursts has matured considerably since their discovery over thirty years ago. The determination of a cosmological origin for the long-duration bursts has focused attention on models in which GRBs arise from the death of massive stars, and this has now been confirmed by several lines

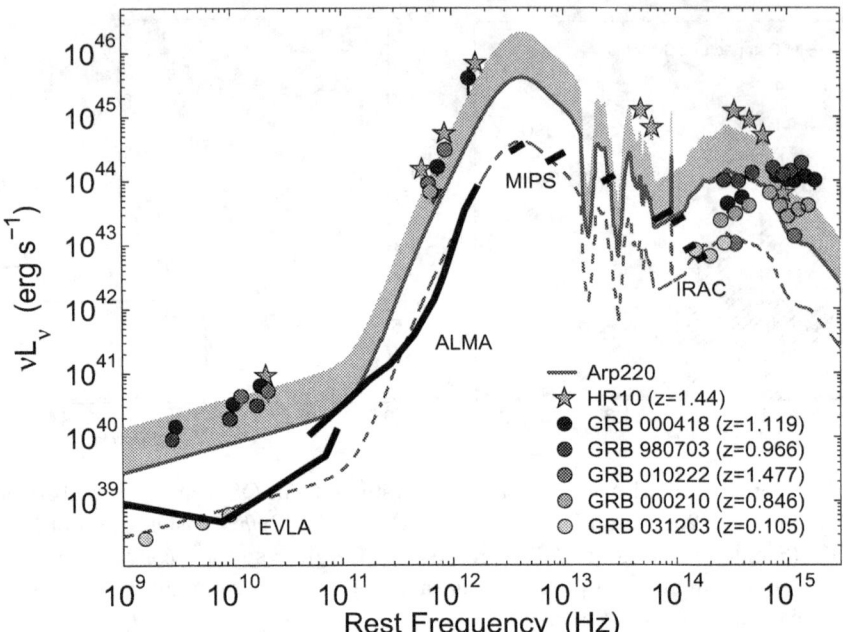

Figure 12.4: Broad-band spectral energy distributions for several GRB hosts detected in the optical/near-IR, submillimeter and radio bands. Also shown are the SED of the local starburst galaxy Arp 220. The thick black lines represent the 300-second sensitivity of the MIPS and IRAC instruments on-board the *Spitzer* space telescope, as well as the projected sensitivity of ALMA and the Expanded VLA. Clearly, these will extend the depth of host galaxy observations by at least an order of magnitude, allowing the detection of systems with moderate luminosities $\sim$ few $\times 10^{10}$ $L_\odot$.

of reasoning, most importantly, the spectroscopic detection of a supernova in association with a GRB. This realization has propelled the field of GRB astronomy in two directions, namely an investigation of the energy source giving rise to the explosion partly in the context of supernovae, and the use of GRBs as tools for the study of star formation and the metal enrichment history of the universe.

The studies presented in this thesis provide quantitative constrains on GRB engine models:

- Gamma-ray burst outflows are narrowly collimated with a wide distribution of jet opening angles ($\sim 5 - 30°$). The jets appear to maintain a simple geometry over a wide range of radii, with a double-jet structure in some cases. This argues against structured jets.

- There is a strong correlation between the energy per unit solid angle and the jet opening angles such that the total relativistic energy release is strongly clustered for GRBs, XRFs, and perhaps SN 1998bw-like events. Values range from about 5×10^{50} to 5×10^{51} erg.

- Type Ibc supernovae in the local universe are primarily powered by the quasi-spherical explosive ejection of the progenitor envelope. Less than 3% are powered by engines.

Whether there exists a population of intermediate sources bridging the supernova and cosmological GRB populations remains an open question.

Regardless of the detailed physics, the extreme luminosity and association with massive stars and star-forming galaxies, makes GRBs a powerful probe of cosmology. Initial studies conducted in the optical/near-IR, submillimeter and radio indicate that GRBs preferentially arise in young starburst galaxies, some of which exhibit extremely large star formation rates. The advent of *Swift* promises a sample of several hundred GRB hosts, some of which with unparalleled information on the metallicity and dynamics of the interstellar medium and intergalactic medium. Combined with advances in far-IR, millimeter and radio facilities, such as *Spitzer*, the Atacama Large Millimeter Array, the Expanded Very Large Array, and the Square Kilometer Array (Figure 12.4), GRBs are poised to make an impact in the quest for understanding the formation and evolution of galaxies and the intergalactic medium.

CHAPTER A

Additional Publications

Refereed Publications Related to Gamma-Ray Bursts

1. Berger, E., Soderberg, A. M., Frail, D. A., Kulkarni, S. R. 2003, ApJ, 587, L5
   A Radio Flare from GRB 020405: Evidence for a Uniform Medium Around a Massive Stellar Progenitor

2. Lipkin, Y. M., Ofek, E. O., Gal-Yam, A., Leibowitz, E. M., Poznanski, D., Kaspi, S., Polishook, D., Kulkarni, S. R., Fox, D. W., Berger, E., Mirabal, N., Halpern, J., Bureau, M., Fathi, K., Price, P. A., Peterson, B. A., Frebel, A., Schmidt, B., Orosz, J. A., Fitzgerald, J. B., Bloom, J. S., van Dokkum, P. G., Bailyn, C. D., Buxton, M. M., Barsony, M. 2004, ApJ, 606, 381
   The Detailed Optical Light Curve of GRB 030329

3. Frail, D. A., Metzger, B. D., Berger, E., Kulkarni, S. R., Yost, S. A. 2004, ApJ, 600, 828
   A Late-Time Flattening of Afterglow Light Curves

4. Bloom, J. S., Fox, D., van Dokkum, P. G., Kulkarni, S. R., Berger, E., Djorgovski, S. G., Frail, D. A. 2003, ApJ, 599, 957
   The First Two Host Galaxies of X-Ray Flashes: XRF 011030 and XRF 020427

5. Sheth, K., Frail, D. A., White, S., Das, M., Bertoldi, F., Walter, F., Kulkarni, S. R., **Berger, E.**. 2003, ApJ, 595, L33
   Millimeter Observations of GRB 030329: Continued Evidence for a Two-Component Jet

6. Price, P. A., Fox, D. W., Kulkarni, S. R., Peterson, B. A., Schmidt, B. P., Soderberg, A. M., Yost, S. A., **Berger, E.**, Djorgovski, S. G., Frail, D. A., Harrison, F. A., Sari, R., Blain, A. W., Chapman, S. C. 2003, Nature, 423, 844
   The bright optical afterglow of the nearby γ-ray burst of 29 March 2003

7. Frail, D. A., Kulkarni, S. R., **Berger, E.**, & Wieringa, M. H. 2003, AJ, 125, 2299
   A Complete Catalog of Radio Afterglows: The First Five Years

8. Li, W., Filippenko, A. V., Chornock, R., **Berger, E.**, Berlind, P., Calkins, M., Challis, P., Fassnacht, C., Jha, S., Kirshner, R., Matheson, T., Sargent, W., Simcoe, R., Smith, G., Squires, G. 2003, PASP, 115, 453
   SN 2002cx: The Most Peculiar Known Type Ia Supernova

9. Frail, D. A., Yost, S. A., **Berger, E.**, Harrison, F. A., Sari, R., Kulkarni, S. R., Taylor, G. B., Bloom, J. S., Fox, D. W., Moriarty-Schieven, G. H., Price, P. A. 2003, ApJ, 590, 992
   The Broadband Afterglow of GRB980703

10. Galama, T. J., Reichart, D., Brown, T. M., Kimble, R. A., Price, P. A., **Berger, E.** Frail, D. A., Kulkarni, S. R., Yost, S. A., Gal-Yam, A., Bloom, J. S., Harrison, F. A., Sari, R., Fox, D., Djorgovski, S. G. 2003, ApJ, 587, 135
    Hubble Space Telescope and Ground-based Optical and Ultraviolet Observations of GRB 010222

11. Galama, T. J., Frail, D. A., Sari, R., **Berger, E.**, Taylor, G. B., & Kulkarni, S. R. 2003, ApJ, 585, 899
    Continued Radio Monitoring of the Gamma-Ray Burst 991208

12. Fox, D. W., Price, P. A., Soderberg, A. M., **Berger, E.**, Kulkarni, S. R., Sari, R., Frail, D. A., Harrison, F. A., Yost, S. A., Matthews, K., Peterson, B. A., Tanaka, I., Christiansen, J., Moriarty-Schieven, G. H. 2003, ApJ, 586, L5
    Discovery of Early Optical Emission from GRB 021211

13. Fox, D. W., Yost, S., Kulkarni, S. R., Torii, K., Kato, T., Yamaoka, H., Sako, M., Harrison, F. A., Sari, R., Price, P. A., **Berger, E.**, Soderberg, A. M., Djorgovski, S. G., Barth, A. J., Pravdo, S. H., Frail, D. A., Gal-Yam, A., Lipkin, Y., Mauch, T., Harrison, C., Buttery, H. 2003, Nature, 422, 284
    Early optical emission from the gamma-ray burst of 4 October 2002

14. Bloom, J. S., **Berger, E.**, Kulkarni, S. R., Djorgovski, S. G., & Frail, D. A. 2003, AJ, 125, 999
    The redshift determination of GRB 990506 and GRB 000418 with the Echellete Spectrograph Imager on Keck

15. Price, P. A., Kulkarni, S R., **Berger, E.**, Fox, D. W., Bloom, J. S., Djorgovski, S. G., Frail, D. A., Galama, T. J., Harrison, F. A., McCarthy, P., Reichart, D. E., Sari, R., Yost, S. A., Jerjen, H., Flint, K., Phillips, A., Warren, B. E., Axelrod, T. S., Chevalier, R. A., Holtzman, J., Kimble, R. A., Schmidt, B. P., Wheeler, J. C., Frontera, F., Costa, E., Piro, L., Hurley, K., Cline, T., Guidorzi, C., Montanari, E., Mazets, E., Golenetskii, S., Mitrofanov, I., Anfimov, D., Kozyrev, A., Litvak, M., Sanin, A., Boynton, W., Fellows, C., Harshman, K., Shinohara, C., Gal-Yam, A., Ofek, E., Lipkin, Y. 2003, ApJ, 589, 838
    Discovery of GRB 020405 and its Late Red Bump

16. Price, P. A., Kulkarni, S. R., Schmidt, B. P., Galama, T. J., Bloom, J. S., **Berger, E.**, Frail, D. A., Djorgovski, S. G., Fox, D. W., Henden, A. A., Klose, S., Harrison, F. A., Reichart, D. E., Sari, R., Yost, S. A., Axelrod, T. S., McCarthy, P., Holtzman, J., Halpern, J. P., Kimble, R. A., Wheeler, J. C., Chevalier, R. A., Hurley, K., Ricker, G. R., Costa, E., Frontera, F., Piro, L. 2003, ApJ, 584, 931
    GRB 010921: Strong Limits on an Underlying Supernova from the Hubble Space Telescope

17. Yost, S. A., Frail, D. A., Harrison, F. A., Sari, R., Reichart, D., Bloom, J. S., Kulkarni, S. R., Moriarty-Schieven, G. H., Djorgovski, S. G., Price, P. A., Goodrich, R. W., Larkin, J. E., Walter, F., Shepherd, D. S., Fox, D. W., Taylor, G. B., **Berger, E.**, Galama, T. J. 2002, ApJ, 577, 155
    The Broadband Afterglow of GRB 980329

18. Price, P. A., **Berger, E.**, Reichart, D., Kulkarni, S. R., Yost, S. A., Subrahmanyan, R., Wark, R. M., Wieringa, M. H., Frail, D. A., Bailey, J., Boyle, B., Corbett, E., Gunn, K., Ryder, S. D., Seymour, N., Koviak, K., McCarthy, P., Phillips, M., Axelrod, T. S., Bloom, J. S., Djorgovski, S. G., Fox, D. W., Galama, T. J., Harrison, F. A., Hurley, K., Sari, R., Schmidt, B. P., Brown, M. J. I., Cline, T., Frontera, F., Guidorzi, C., Montanari, E. 2002, ApJ, 572, L51
    GRB 011121: A Massive Star Progenitor

19. Bloom, J. S., Kulkarni, S. R., Price, P. A., Reichart, D., Galama, T. J., Schmidt, B. P., Frail, D. A., **Berger, E.**, McCarthy, P. J., Chevalier, R. A., Wheeler, J. C., Halpern, J. P., Fox, D. W.,

Djorgovski, S. G., Harrison, F. A., Sari, R., Axelrod, T. S., Kimble, R. A., Holtzman, J., Hurley, K., Frontera, F., Piro, L., Costa, E. 2002, ApJ, 572, L45
Detection of a supernova signature associated with GRB 011121

20. Price, P. A., Kulkarni, S. R., **Berger, E.**, Djorgovski, S. G., Frail, D. A., Mahabal, A., Fox, D. W., Harrison, F. A., Bloom, J. S., Yost, S. A., Reichart, D. E., Henden, A. A., Ricker, G. R., van der Spek, R., Hurley, K., Atteia, J.-L., Kawai, N., Fenimore, E., Graziani, C. 2002, ApJ, 571, L121
GRB 010921: Discovery of the First High Energy Transient Explorer Afterglow

21. Piro, L., Frail, D. A., Gorosabel, J., Garmire, G., Soffitta, P., Amati, L., Andersen, M. I., Antonelli, L. A., **Berger, E.**, Frontera, F., Fynbo, J., Gandolfi, G., Garcia, M. R., Hjorth, J., Zand, J. i., Jensen, B. L., Masetti, N., Møller, P., Pedersen, H., Pian, E., Wieringa, M. H. 2002, ApJ, 577, 680
The Bright Gamma-Ray Burst of February 10, 2000: A Case Study of an Optically Dark GRB

22. Price, P. A., **Berger, E.**, Kulkarni, S. R., Djorgovski, S. G., Fox, D. W., Mahabal, A., Hurley, K., Bloom, J. S., Frail, D. A., Galama, T. J., Harrison, F. A., Morrison, G., Reichart, D. E., Yost, S. A., Sari, R., Axelrod, T. S., Cline, T., Golenetskii, S., Mazets, E., Schmidt, B. P., Trombka, J. 2002, ApJ, 573, 85
The Unusually Long Duration Gamma-ray Burst GRB 000911: Discovery of the Afterglow and Host Galaxy

23. Lazzati, D., Covino, S., Ghisellini, G., Fugazza, D., Campana, S., Saracco, P., Price, P. A., **Berger, E.**, Kulkarni, S., Ramirez-Ruiz, E., Cimatti, A., Della Valle, M., di Serego Alighieri, S., Celotti, A., Haardt, F., Israel, G. L., Stella, L. 2001, A&A, 378, 996
The optical afterglow of GRB 000911: evidence for an associated supernova?

24. Frail, D. A., Bertoldi, F., Moriarty-Schieven, G. H., **Berger, E.**, Price, P. A., Bloom, J. S., Sari, R., Kulkarni, S. R., Gerardy, C. L., Reichart, D. E., Djorgovski, S. G., Galama, T. J., Harrison, F. A., Walter, F., Shepherd, D. S., Halpern, J., Peck, A. B., Menten, K. M., Yost, S. A., Fox, D. W. 2002, ApJ, 565, 829
GRB 010222: a Burst Within a Starburst

25. Hurley, K., **Berger, E.**, Castro-Tirado, A., Castro Cerón, J. M., Cline, T., Feroci, M., Frail, D. A., Frontera, F., Masetti, N., Guidorzi, C., Montanari, E., Hartmann, D. H., Henden, A., Levine, S. E., Mazets, E., Golenetskii, S., Frederiks, D., Morrison, G., Oksanen, A., Moilanen, M., Park, H.-S., Price, P. A., Prochaska, J., Trombka, J., Williams, G. 2002, ApJ, 567, 447
Afterglow upper limits for four short duration, hard spectrum gamma-ray bursts

26. Harrison, F. A., Yost, S. A., Sari, R., **Berger, E.**, Galama, T. J., Holtzman, J., Axelrod, T., Bloom, J. S., Chevalier, R., Costa, E., Diercks, A., Djorgovski, S. G., Frail, D. A., Frontera, F., Hurley, K., Kulkarni, S. R., McCarthy, P., Piro, L., Pooley, G. G., Price, P. A., Reichart, D., Ricker, G. R., Shepherd, D., Schmidt, B., Walter, F., Wheeler, C. 2001, ApJ, 559, 123
Broadband Observations of the Afterglow of GRB 000926: Observing the Effect of Inverse Compton Scattering

27. Frail, D. A., Kulkarni, S. R., Sari, R., Djorgovski, S. G., Bloom, J. S., Galama, T. J., Reichart, D. E., **Berger, E.**, Harrison, F. A., Price, P. A., Yost, S. A., Diercks, A., Goodrich, R. W., Chaffee, F. 2001, ApJ, 562, L55
Beaming in Gamma-Ray Bursts: Evidence for a Standard Energy Reservoir

28. Price, P. A., Harrison, F. A., Galama, T. J., Reichart, D. E., Axelrod, T. S., **Berger, E.**, Bloom, J. S., Busche, J., Cline, T., Diercks, A., Djorgovski, S. G., Frail, D. A., Gal-Yam, A., Halpern,

J., Holtzman, J. A., Hunt, M., Hurley, K., Jacoby, B., Kimble, R., Kulkarni, S. R., Mirabal, N., Morrison, G., Ofek, E., Pevunova, O., Sari, R., Schmidt, B. P., Turnshek, D., Yost, S. 2001, ApJ, 549, L7
Multi-Color Observations of the GRB 000926 Afterglow

29. Galama, T. J., Bremer, M., Bertoldi, F., Menten, K. M., Lisenfeld, U., Shepherd, D. S., Mason, B., Walter, F., Pooley, G. G., Frail, D. A., Sari, R., Kulkarni, S. R., **Berger, E.**, Bloom, J. S., Castro-Tirado, A. J., Granot, J. 2000, ApJ, 541, L45
The Bright Gamma-Ray Burst 991208 — Tight Constraints on Afterglow Models from Observations of the Early-Time Radio Evolution

30. Frail, D. A., **Berger, E.**, Galama, T. J., Kulkarni, S. R., Moriarty-Schieven, G. H., Pooley, G. G., Sari, R., Shepherd, D. S., Taylor, G. B., Walter, F. 2000, ApJ, 538, L129
The Enigmatic Radio Afterglow of GRB 991216

SECTION A.2
Refereed Publications not Related to Gamma-Ray Bursts

---

1. Berger, E. 2002, ApJ, 572, 503
Flaring Up All Over: Radio Activity in Rapidly-Rotating Late-Type M and L Dwarfs

2. Berger, E., Ball, S., Becker, K. M., Clarke, M., Frail, D. A., Fukuda, T. A., Hoffman, I. M., Mellon, R., Momjian, E., Murphy, N. W., Teng, S. H., Woodruff, T., Zauderer, B. A., Zavala, R. T. 2001, Nature, 410, 338
Discovery of Radio Emission from the Brown Dwarf LP944-20

SECTION A.3
Conference Proceedings

---

1. Berger, E. 2003, to appear in proceeding of "3-D signatures in stellar explosions", June 10-13 2003, Austin, Texas, USA; astro-ph/0309714
The Diversity of Cosmic Explosions: Gamma-Ray Bursts and Type Ib/c Supernovae

2. Berger, E. 2003, to appear in proceedings of "IAU Colloquium 192, Supernovae: 10 Years of 1993J", Valencia, Spain 22-26 April 2003, eds. J. M. Marcaide, K. W. Weiler; astro-ph/0309713
How Common are Engines in Ib/c Supernovae?

3. Berger, E., Kulkarni, S. R., & Frail, D. A. 2002, American Astronomical Society, 201st AAS Meeting, #84.03; Bulletin of the American Astronomical Society, Vol. 34, p.1243
Calorimetry of GRB afterglows – a Direct Measure of the Fireball Energy

4. Berger, E. 2002, to appear in proceedings of the "Gamma-Ray Bursts in the Afterglow Era: 3rd Workshop", ASPCS, in press; astro-ph/0304346
Radio Afterglows of Gamma-Ray Bursts: Unique Clues to the Energetics and Environments

5. Berger, E. 2001, to appear in Proc. of the Gamma-Ray Burst and Afterglow Astronomy 2001: A Workshop Celebrating the First Year of the HETE Mission; astro-ph/0112559
Radio, Sub-mm, and X-ray Studies of Gamma-Ray Burst Host Galaxies

6. Berger, E., Ball, S., Becker, K. M., et al. 2001, American Astronomical Society, 198th AAS Meeting, #69.06; Bulletin of the American Astronomical Society, Vol. 33, p.891
Discovery of Radio Emission from the Brown Dwarf LP944-20 and Preliminary Results from an Ongoing Survey of Nearby Brown Dwarfs with the VLA

7. Berger, E., Sari, R., Frail, D. A., Kulkarni, S. R. Gamma-Ray Bursts in the Afterglow Era: Rome Workshop. Edited by E. Costa, F. Frontera, and J. Hjorth. Berlin Heidelberg: Springer, 2001, p. 154.
   Broad-band Modeling of GRB Afterglow

Bibliography

Adelberger, K. L. and Steidel, C. C. 2000, ApJ, 544, 218

Adelberger, K. L., Steidel, C. C., Shapley, A. E., and Pettini, M. 2003, ApJ, 584, 45

Akerlof, C. *et al.* . 1999, Nature, 398, 400

Alard, C. 2000, A&AS, 144, 363

—. 2001, A&A, 377, 389

Amati, L. *et al.* . 2000a, GRB Circular Network, 885, 1

—. 2002, A&A, 390, 81

—. 2000b, Science, 290, 953

Andersen, M. I. *et al.* . 2000, A&A, 364, L54

Antonelli, L. A. *et al.* . 1999, A&AS, 138, 435

—. 2000, GRB Circular Network, 561, 1

Baird, G. A. *et al.* . 1975, ApJ, 196, L11

Baird, G. A., Meikle, W. P. S., Jelley, J. V., Palumbo, G. G. C., and Partridge, R. B. 1976, A&SS, 42, 69

Baker, A. J., Lutz, D., Genzel, R., Tacconi, L. J., and Lehnert, M. D. 2001, A&A, 372, L37

Barat, C., Chambon, G., Hurley, K., Niel, M., Vedrenne, G., Estulin, I. V., Kuznetsov, A. V., and Zenchenko, V. M. 1981, Space Science Instrumentation, 5, 229

Barger, A. J., Cowie, L. L., Mushotzky, R. F., and Richards, E. A. 2001, AJ, 121, 662

Barger, A. J., Cowie, L. L., and Richards, E. A. 2000, AJ, 119, 2092

Barger, A. J., Cowie, L. L., and Sanders, D. B. 1999, ApJ, 518, L5

Barkana, R. and Loeb, A. 2001, Phys. Rep., 349, 125

Barnard, V. E. *et al.* . 2003, MNRAS, 338, 1

Baron, E., Young, T. R., and Branch, D. 1993, ApJ, 409, 417

Becker, R. H. *et al.* . 2001, AJ, 122, 2850

Berger, E. *et al.* . 2001a, ApJ, 556, 556

—. 2002a, ApJ, 581, 981

Berger, E., Kulkarni, S. R., and Chevalier, R. A. 2002b, ApJ, 577, L5

Berger, E., Kulkarni, S. R., and Frail, D. A. 2001b, ApJ, 560, 652

Berger, E., Kulkarni, S. R., and Frail, D. A. 2002c, in International Astronomical Union Circular, 2–+

—. 2003a, ApJ, 590, 379

Berger, E., Kulkarni, S. R., Frail, D. A., and Soderberg, A. M. 2003b, ApJ, 599, 408

Berger, E. *et al.* . 2003c, Nature, 426, 154

—. 2000, ApJ, 545, 56

Berger, E., Soderberg, A. M., Frail, D. A., and Kulkarni, S. R. 2003d, ApJ, 587, L5

Bertoldi, F. 2000, GRB Circular Network, 580, 1

Bessell, M. S. and Brett, J. M. 1988, PASP, 100, 1134

Bessell, M. S. and Germany, L. M. 1999, PASP, 111, 1421

Bisnovatyi-Kogan, G. S., Imshennik, V. S., Nadiozhin, D. K., and Chechetkin, V. M. 1975, A&SS, 35, 23

Blain, A. W. and Natarajan, P. 2000, MNRAS, 312, L35

Blain, A. W., Smail, I., Ivison, R. J., Kneib, J.-P., and Frayer, D. T. 2002, Phys. Rep., 369, 111

Blandford, R. D. and McKee, C. F. 1976, Physics of Fluids, 19, 1130

Bloom, J. S. 2002, GRB Circular Network, 1225, 1

Bloom, J. S., Berger, E., Kulkarni, S. R., Djorgovski, S. G., and Frail, D. A. 2003a, AJ, 125, 999

Bloom, J. S., Diercks, A., Djorgovski, S. G., Kaplan, D., and Kulkarni, S. R. 2000, GRB Circular Network, 661, 1

Bloom, J. S., Frail, D. A., and Kulkarni, S. R. 2003b, ApJ, 594, 674

Bloom, J. S. *et al.* . 1998a, ApJ, 508, L21

Bloom, J. S., Frail, D. A., and Sari, R. 2001, AJ, 121, 2879

Bloom, J. S., Kulkarni, S. R., and Djorgovski, S. G. 2002a, AJ, 123, 1111

Bloom, J. S. *et al.* . 1999, Nature, 401, 453

Bloom, J. S., Kulkarni, S. R., Harrison, F., Prince, T., Phinney, E. S., and Frail, D. A. 1998b, ApJ, 506, L105

Bloom, J. S. *et al.* . 2002b, ApJ, 572, L45

Bloom, J. S., Price, P. A., Fox, D., and Kulkarni, S. R. 2002c, GRB Circular Network, 1389, 1

Boella, G., Butler, R. C., Perola, G. C., Piro, L., Scarsi, L., and Bleeker, J. A. M. 1997, A&AS, 122, 299

Branch, D., Nomoto, K., and Filippenko, A. V. 1990, Comments on Astrophysics, 15, 221

Cappellaro, E., Evans, R., and Turatto, M. 1999, A&A, 351, 459

Cardelli, J. A., Clayton, G. C., and Mathis, J. S. 1989, ApJ, 345, 245

Carilli, C. L. and Yun, M. S. 1999, ApJ, 513, L13

—. 2000, ApJ, 530, 618

Castro, S. M., Diercks, A., Djorgovski, S. G., Kulkarni, S. R., Galama, T. J., Bloom, J. S., Harrison, F. A., and Frail, D. A. 2000, GRB Circular Network, 605, 1

Castro-Tirado, A. J. *et al.* . 2001, A&A, 370, 398

—. 1999, ApJ, 511, L85

Cavallo, G. and Rees, M. J. 1978, MNRAS, 183, 359

Chapman, S. C., Blain, A. W., Ivison, R. J., and Smail, I. R. 2003, Nature, 422, 695

Chapman, S. C., Lewis, G. F., Scott, D., Borys, C., and Richards, E. 2002a, ApJ, 570, 557

Chapman, S. C., Richards, E. A., Lewis, G. F., Wilson, G., and Barger, A. J. 2001, ApJ, 548, L147

Chapman, S. C. *et al.* . 2000, MNRAS, 319, 318

Chapman, S. C., Shapley, A., Steidel, C., and Windhorst, R. 2002b, ApJ, 572, L1

Chary, R., Becklin, E. E., and Armus, L. 2002, ApJ, 566, 229

Chevalier, R. A. 1982, ApJ, 258, 790

—. 1984, ApJ, 285, L63

—. 1998, ApJ, 499, 810

Chevalier, R. A. and Li, Z. 1999, ApJ, 520, L29

—. 2000, ApJ, 536, 195

Chornock, R. 2002, in International Astronomical Union Circular, 2–+

Chornock, R. and Filippenko, A. V. 2001, in International Astronomical Union Circular, 2–+

Cline, T. L. and Desai, U. D. 1976, A&SS, 42, 17

Cline, T. L., Desai, U. D., Klebesadel, R. W., and Strong, I. B. 1973, ApJ, 185, L1+

Clocchiatti, A. *et al.* . 2000, ApJ, 529, 661

Cohen, J. G., Hogg, D. W., Blandford, R., Cowie, L. L., Hu, E., Songaila, A., Shopbell, P., and Richberg, K. 2000, ApJ, 538, 29

Colgate, S. A. 1968, Canadian Journal of Physics, 46, 476

Condon, J. J. 1992, ARAA, 30, 575

Condon, J. J., Helou, G., and Jarrett, T. H. 2002, AJ, 123, 1881

Condon, J. J., Helou, G., Sanders, D. B., and Soifer, B. T. 1993, AJ, 105, 1730

Cortiglioni, S., Mandolesi, N., Morigi, G., Ciapi, A., Inzani, P., and Sironi, G. 1981, A&SS, 75, 153

Costa, E. *et al.* . 1997, Nature, 387, 783

Curran, S. J., Webb, J. K., Murphy, M. T., Bandiera, R., Corbelli, E., and Flambaum, V. V. 2002, Publications of the Astronomical Society of Australia, 19, 455

Dai, Z. G. and Lu, T. 2001, A&A, 367, 501

De Pasquale, M. *et al.* . 2002, submitted to ApJ; astro-ph/0212298

Dermer, C. D. and Mitman, K. E. 1999, ApJ, 513, L5

Dey, A., Graham, J. R., Ivison, R. J., Smail, I., Wright, G. S., and Liu, M. C. 1999, ApJ, 519, 610

Djorgovski, S. G., Frail, D. A., Kulkarni, S. R., Bloom, J. S., Odewahn, S. C., and Diercks, A. 2001a, ApJ, 562, 654

Djorgovski, S. G. *et al.* . 2001b, in Gamma-ray Bursts in the Afterglow Era, 218–+

Djorgovski, S. G., Kulkarni, S. R., Bloom, J. S., Goodrich, R., Frail, D. A., Piro, L., and Palazzi, E. 1998, ApJ, 508, L17

Dressler, A. and Gunn, J. E. 1982, ApJ, 263, 533

Dunlop, J. S. *et al.* . 2002, ArXiv Astrophysics e-prints

Dunne, L., Clements, D. L., and Eales, S. A. 2000, MNRAS, 319, 813

Eichler, D., Livio, M., Piran, T., and Schramm, D. N. 1989, Nature, 340, 126

Elbaz, D., Flores, H., Chanial, P., Mirabel, I. F., Sanders, D., Duc, P.-A., Cesarsky, C. J., and Aussel, H. 2002, A&A, 381, L1

Falcke, H., Lehar, J., Barvainis, R., Nagar, N. M., and Wilson, A. S. 2001, in ASP Conf. Ser. 224: Probing the Physics of Active Galactic Nuclei, 265–+

Feroci, M. *et al.* . 1998, A&A, 332, L29

—. 2001, A&A, 378, 441

—. 2000, GRB Circular Network, 685, 1

Filippenko, A. V. 1997, ARAA, 35, 309

Filippenko, A. V. and Chornock, R. 2001, in International Astronomical Union Circular, 1–+

Fishman, G. J. and Meegan, C. A. 1995, ARAA, 33, 415

Fitzpatrick, E. L. and Massa, D. 1988, ApJ, 328, 734

Folkes, S. *et al.* . 1999, MNRAS, 308, 459

Fomalont, E. 1981, NEWSLETTER. NRAO NO. 3, P. 3, 1981, 3, 3

Fomalont, E. B., Kellermann, K. I., Partridge, R. B., Windhorst, R. A., and Richards, E. A. 2002, AJ, 123, 2402

Fomalont, E. B., Windhorst, R. A., Kristian, J. A., and Kellerman, K. I. 1991, AJ, 102, 1258

Fox, D. W. *et al.* . 2003, Nature, 422, 284

Frail, D. A. *et al.* . 2000a, ApJ, 538, L129

—. 2002, ApJ, 565, 829

Frail, D. A., Goss, W. M., and Whiteoak, J. B. Z. 1994, ApJ, 437, 781

Frail, D. A. and Kulkarni, S. R. 1995, A&SS, 231, 277

Frail, D. A., Kulkarni, S. R., Berger, E., and Wieringa, M. H. 2003a, AJ, 125, 2299

Frail, D. A., Kulkarni, S. R., Nicastro, S. R., Feroci, M., and Taylor, G. B. 1997, Nature, 389, 261

Frail, D. A. *et al.* . 2001, ApJ, 562, L55

—. 2000b, ApJ, 534, 559

Frail, D. A., Metzger, B. D., Berger, E., Kulkarni, S. R., and Yost, S. A. 2004, ApJ, 600, 828

Frail, D. A., Waxman, E., and Kulkarni, S. R. 2000c, ApJ, 537, 191

Frail, D. A. *et al.* . 2003b, ApJ, 590, 992

Frayer, D. T. *et al.* . 1999, ApJ, 514, L13

Frayer, D. T., Ivison, R. J., Scoville, N. Z., Yun, M., Evans, A. S., Smail, I., Blain, A. W., and Kneib, J.-P. 1998, ApJ, 506, L7

Freedman, D. L. and Waxman, E. 2001, ApJ, 547, 922

Frontera, F. *et al.* . 2000, ApJ, 540, 697

—. 1999, GRB Circular Network, 401, 1

—. 1998, ApJ, 493, L67

—. 2001, GRB Circular Network, 950, 1

Fruchter, A. S. and Hook, R. N. 2002, PASP, 114, 144

Fukugita, M., Shimasaku, K., and Ichikawa, T. 1995, PASP, 107, 945

Fynbo, J. P. U. *et al.* . 2003, A&A, 406, L63

Fynbo, J. U. *et al.* . 2001, A&A, 369, 373

Galama, T. J. *et al.* . 2000, ApJ, 541, L45

Galama, T. J., Frail, D. A., Sari, R., Berger, E., Taylor, G. B., and Kulkarni, S. R. 2003, ApJ, 585, 899

Galama, T. J. *et al.* . 1998a, Nature, 395, 670

—. 1998b, Nature, 395, 670

—. 1998c, Nature, 395, 670

Galama, T. J. and Wijers, R. A. M. J. 2001, ApJ, 549, L209

Galama, T. J., Wijers, R. A. M. J., Bremer, M., Groot, P. J., Strom, R. G., Kouveliotou, C., and van Paradijs, J. 1998d, ApJ, 500, L97+

Ganeshalingam, M. and Li, W. D. 2002, in International Astronomical Union Circular, 1–+

Garcia, M. R. *et al.* . 1998, ApJ, 500, L105+

Garnavich, P., Jha, S., Kirshner, R., and Berlind, P. 1998, in International Astronomical Union Circular, 1–+

Germany, L. M., Reiss, D. J., Sadler, E. M., Schmidt, B. P., and Stubbs, C. W. 2000, ApJ, 533, 320

Goodman, J. 1986, ApJ, 308, L47

—. 1997, New Astronomy, 2, 449

Gorenstein, P., Helmken, H., and Gursky, H. 1976, A&SS, 42, 89

Granot, J. and Loeb, A. 2003, ApJ, 593, L81

Granot, J., Nakar, E., and Piran, T. 2003, ArXiv Astrophysics e-prints, 4563

Granot, J., Panaitescu, A., Kumar, P., and Woosley, S. E. 2002, ApJ, 570, L61

Granot, J., Piran, T., and Sari, R. 1999a, ApJ, 513, 679

—. 1999b, ApJ, 527, 236

Granot, J. and Sari, R. 2002, ApJ, 568, 820

Greiner, J., Flohrer, J., Wenzel, W., and Lehmann, T. 1987, A&SS, 138, 155

Grindlay, J. E., Wright, E. L., and McCrosky, R. E. 1974, ApJ, 192, L113+

Groot, P. J. *et al.* . 1998, ApJ, 502, L123+

Höflich, P., Wheeler, J. C., and Wang, L. 1999, ApJ, 521, 179

Haarsma, D. B., Partridge, R. B., Windhorst, R. A., and Richards, E. A. 2000, ApJ, 544, 641

Halpern, J. P. *et al.* . 2000, ApJ, 543, 697

Hamuy, M. 2003, ApJ, 582, 905

Hanlon, L. *et al.* . 2000, A&A, 359, 941

Harrison, F. A. *et al.* . 1999, ApJ, 523, L121

—. 2001, ApJ, 559, 123

Hartmann, D. and Epstein, R. I. 1989, ApJ, 346, 960

Harwit, M. and Salpeter, E. E. 1973, ApJ, 186, L37+

Heise, J., in 't Zand, J. J. M., and Kulkarni, S. R. 2003, in prep.

Helou, G., Soifer, B. T., and Rowan-Robinson, M. 1985, ApJ, 298, L7

Henden, A. 2000, GRB Circular Network, 662, 1

—. 2002, GRB Circular Network, 1251, 1

Henden, A., Canzian, B., Zeh, A., and Klose, S. 2003, GRB Circular Network, 2123, 1

Henden, A., Harris, H., and Klose, S. 2000, GRB Circular Network, 652, 1

Hjorth, J. and et al. 2003, Nature in press

Hjorth, J. *et al.* . 2003, Nature, 423, 847

—. 2002, ApJ, 576, 113

Ho, C., Epstein, R. I., and Fenimore, E. E. 1990, ApJ, 348, L25

Holland, S. *et al.* . 2001, A&A, 371, 52

Hopkins, A. M., Connolly, A. J., Haarsma, D. B., and Cram, L. E. 2001, AJ, 122, 288

Hu, E. M. and Ridgway, S. E. 1994, AJ, 107, 1303

Hudec, R. *et al.* . 1987, A&A, 175, 71

Hughes, D. H. *et al.* . 1998, Nature, 394, 241

Hurley, K., Cline, T., and Mazets, E. 2000, GRB Circular Network, 642, 1

Hurley, K. *et al.* . 2002, GRB Circular Network, 1223, 1

Ibrahimov, M. A., Asfandiyarov, I. M., Kahharov, B. B., Pozanenko, A., Rumyantsev, V., and Beskin, G. 2003, GRB Circular Network, 2191, 1

Imhof, W. L., Nakano, G. H., Johnson, R. G., Kilner, J. R., Regan, J. B., Klebesadel, R. W., and Strong, I. B. 1974, ApJ, 191, L7+

in 't Zand, J. J. M. *et al.* . 1998, ApJ, 505, L119

in't Zand, J. J. M. *et al.* . 2001, ApJ, 559, 710

Ivison, R. J., Smail, I., Barger, A. J., Kneib, J.-P., Blain, A. W., Owen, F. N., Kerr, T. H., and Cowie, L. L. 2000, MNRAS, 315, 209

Ivison, R. J., Smail, I., Le Borgne, J.-F., Blain, A. W., Kneib, J.-P., Bezecourt, J., Kerr, T. H., and Davies, J. K. 1998, MNRAS, 298, 583

Iwamoto, K. *et al.* . 1998, Nature, 395, 672

—. 2000, ApJ, 534, 660

Iwamoto, K., Nomoto, K., Hoflich, P., Yamaoka, H., Kumagai, S., and Shigeyama, T. 1994, ApJ, 437, L115

Jaunsen, A. O. *et al.* . 2001, ApJ, 546, 127

Jorgensen, I. 1994, PASP, 106, 967

Kashikawa, N. *et al.* . 2003, AJ, 125, 53

Kawabata, K. S. *et al.* . 2002, ApJ, 580, L39

Kennicutt, R. C. 1992, ApJ, 388, 310

—. 1998, ARAA, 36, 189

Kinugasa, K., Kawakita, H., Ayani, K., Kawabata, T., and Yamaoka, H. 2002, in International Astronomical Union Circular, 1–+

Klebesadel, R. *et al.* . 1982, ApJ, 259, L51

Klebesadel, R. W., Strong, I. B., and Olson, R. A. 1973, ApJ, 182, L85+

Klose, S. *et al.* . 2000a, GRB Circular Network, 645, 1

—. 2000b, ApJ, 545, 271

Kommers, J. M., Lewin, W. H. G., Kouveliotou, C., van Paradijs, J., Pendleton, G. N., Meegan, C. A., and Fishman, G. J. 2000, ApJ, 533, 696

Kouveliotou, C., Meegan, C. A., Fishman, G. J., Bhat, N. P., Briggs, M. S., Koshut, T. M., Paciesas, W. S., and Pendleton, G. N. 1993, ApJ, 413, L101

Kreysa, E. *et al.* . 1998, in Proc. SPIE Vol. 3357, p. 319-325, Advanced Technology MMW, Radio, and Terahertz Telescopes, Thomas G. Phillips; Ed., 319–325

Krolik, J. H. and Pier, E. A. 1991, ApJ, 373, 277

Kulkarni, S. R. *et al.* . 2000, in Proc. SPIE Vol. 4005, p. 9-21, Discoveries and Research Prospects from 8- to 10-Meter-Class Telescopes, Jacqueline Bergeron; Ed., 9–21

Kulkarni, S. R. *et al.* . 1999a, Nature, 398, 389

—. 1999b, ApJ, 522, L97

—. 1998, Nature, 395, 663

Kumar, P. 2000, ApJ, 538, L125

Kumar, P. and Piran, T. 2000, ApJ, 535, 152

Lamb, D. Q. 1995, PASP, 107, 1152

Lamb, D. Q., Donaghy, T. Q., and Graziani, C. 2004, ApJ (submitted), astro-ph/0312634

Lamb, D. Q. and Reichart, D. E. 2000, ApJ, 536, 1

Lazzati, D., Covino, S., and Ghisellini, G. 2002, MNRAS, 330, 583

Le Floc'h, E. *et al.* . 2003, A&A, 400, 499

Li, H. and Dermer, C. D. 1992, Nature, 359, 514

Li, W., Filippenko, A. V., Chornock, R., and Jha, S. 2003, ApJ, 586, L9

Li, Z. and Chevalier, R. A. 1999, ApJ, 526, 716

—. 2001, ApJ, 551, 940

Lilly, S. J., Le Fevre, O., Hammer, F., and Crampton, D. 1996, ApJ, 460, L1+

Livio, M. and Waxman, E. 2000, ApJ, 538, 187

Lyne, A. G. and Lorimer, D. R. 1994, Nature, 369, 127

Lyutikov, M. and Blandford, R. 2003, ArXiv Astrophysics e-prints

MacFadyen, A. I. and Woosley, S. E. 1999, ApJ, 524, 262

MacFadyen, A. I., Woosley, S. E., and Heger, A. 2001, ApJ, 550, 410

Madau, P., Ferguson, H. C., Dickinson, M. E., Giavalisco, M., Steidel, C. C., and Fruchter, A. 1996, MNRAS, 283, 1388

Marshall, F. E. and Takeshima, T. 1998, GRB Circular Network, 58, 1

Marshall, F. E., Takeshima, T., Kippen, T., and Giblin, R. M. 2000, GRB Circular Network, 519, 1

Marzke, R. O., da Costa, L. N., Pellegrini, P. S., Willmer, C. N. A., and Geller, M. J. 1998, ApJ, 503, 617

Masetti, N. *et al.* . 2000a, A&A, 359, L23

—. 2000b, A&A, 354, 473

Matheson, T., Filippenko, A. V., Li, W., Leonard, D. C., and Shields, J. C. 2001a, AJ, 121, 1648

Matheson, T., Jha, S., Challis, P., Kirshner, R., and Calkins, M. 2001b, in International Astronomical Union Circular, 2–+

Matheson, T., Jha, S., Challis, P., Kirshner, R., Calkins, M., Chornock, R., Li, W. D., and Filippenko, A. V. 2002, in International Astronomical Union Circular, 1–+

Matthews, K. and Soifer, B. T. 1994, Experimental Astronomy, 3, 77

Matzner, C. D. and McKee, C. F. 1999, ApJ, 526, L109

Mazzali, P. A. *et al.* . 2002, ApJ, 572, L61

Mazzali, P. A., Iwamoto, K., and Nomoto, K. 2000, ApJ, 545, 407

McNamara, B. J., Harrison, T. E., and Williams, C. L. 1995, ApJ, 452, L25+

Meegan, C. A., Fishman, G. J., Wilson, R. B., Horack, J. M., Brock, M. N., Paciesas, W. S., Pendleton, G. N., and Kouveliotou, C. 1992, Nature, 355, 143

Meikle, P., Lucy, L., Smartt, S., Leibundgut, B., Lundqvist, P., and Ostensen, R. 2002, in International Astronomical Union Circular, 2–+

Meszaros, P. and Rees, M. J. 1992, MNRAS, 257, 29P

—. 1997, ApJ, 482, L29+

Metzger, A. E., Parker, R. H., Gilman, D., Peterson, L. E., and Trombka, J. I. 1974, ApJ, 194, L19

Metzger, M. R., Djorgovski, S. G., Kulkarni, S. R., Steidel, C. C., Adelberger, K. L., Frail, D. A., Costa, E., and Frontera, F. 1997, Nature, 387, 879

Metzger, M. R. and Fruchter, A. 2000, GRB Circular Network, 669, 1

Meurer, G. R., Heckman, T. M., and Calzetti, D. 1999, ApJ, 521, 64

Millard, J. *et al.* . 1999, ApJ, 527, 746

Mirabal, N., Halpern, J. P., Kemp, J., and Helfand, D. J. 2000, GRB Circular Network, 646, 1

Mirabal, N., Paerels, F., and Halpern, J. P. 2002, ApJ (submitted), astro-ph/0209516

Mitrofanov, I. G. *et al.* . 1991, Soviet Astronomy, 35, 367

Mochkovitch, R., Hernanz, M., Isern, J., and Martin, X. 1993, Nature, 361, 236

Murakami, T., Ueda, Y., Ishida, M., Fujimoto, R., Yoshida, A., and Kawai, N. 1997, IAUC, 6722, 1

Nakamura, T. 1999, ApJ, 522, L101

Nakamura, T., Mazzali, P. A., Nomoto, K., and Iwamoto, K. 2001, ApJ, 550, 991

Nakano, S., Kushida, R., Kushida, Y., and Li, W. 2002, in International Astronomical Union Circular, 1–+

Narayan, R., Paczynski, B., and Piran, T. 1992, ApJ, 395, L83

Nemiroff, R. J. 1994, in AIP Conf. Proc. 307: Gamma-Ray Bursts, 730–+

Nicastro, L. *et al.* . 1999a, A&AS, 138, 437

Nicastro, L., Antonelli, L. A., Dadina, M., Daniele, M. R., Costa, E., and Pian, E. 1999b, IAUC, 7213, 2

Nomoto, K., Yamaoka, H., Pols, O. R., van den Heuvel, E. P. J., Iwamoto, K., Kumagai, S., and Shigeyama, T. 1994, Nature, 371, 227

Norris, J. P. 2002, ApJ, 579, 386

Norris, J. P., Cline, T. L., Desai, U. D., and Teegarden, B. J. 1984, Nature, 308, 434

Oh, S. P. 2002, MNRAS, 336, 1021

O'Mongain, E. and Weekes, T. C. 1974, PASP, 86, 470

Ostrowski, M. and Bednarz, J. 2002, A&A, 394, 1141

Overzier, R. A., Röttgering, H. J. A., Kurk, J. D., and De Breuck, C. 2001, A&A, 367, L5

Paczynski, B. 1986, ApJ, 308, L43

—. 1991a, Acta Astronomica, 41, 257

—. 1991b, Acta Astronomica, 41, 157

—. 1995, PASP, 107, 1167

—. 1998, ApJ, 494, L45+

—. 2001, Acta Astronomica, 51, 1

Paczynski, B. and Xu, G. 1994, ApJ, 427, 708

Panagia, N., Sramek, R. A., and Weiler, K. W. 1986, ApJ, 300, L55

Panaitescu, A. 2001, ApJ, 556, 1002

Panaitescu, A. and Kumar, P. 2000, ApJ, 543, 66

—. 2002, ApJ, 571, 779

Panaitescu, A., Meszaros, P., and Rees, M. J. 1998, ApJ, 503, 314

Patton, D. R. *et al.* . 2002, ApJ, 565, 208

Peacock, J. A. *et al.* . 2000, MNRAS, 318, 535

Pian, E. *et al.* . 2000, ApJ, 536, 778

—. 2001, A&A, 372, 456

Piran, T., Kumar, P., Panaitescu, A., and Piro, L. 2001, ApJ, 560, L167

Piro, L. 2001, in AIP Conf. Proc. 599: X-ray Astronomy: Stellar Endpoints, AGN, and the Diffuse X-ray Background, 295–+

Piro, L. *et al.* . 1998, A&A, 331, L41

—. 2002, ApJ (submitted), astro-ph/0201282

—. 2000, Science, 290, 955

Price, P. A., Bloom, J. S., Goodrich, R. W., Barth, A. J., Cohen, M. H., and Fox, D. W. 2002a, GRB Circular Network, 1475, 1

Price, P. A. and et al. 2002, ApJ (submitted), astro-ph/0208008

Price, P. A. *et al.* . 2003, Nature in press

Price, P. A., Fox, D. W., Yost, S. A., Pravdo, S., Helin, E., Lawrence, K., and Hicks, M. 2002b, GRB Circular Network, 1221, 1

Price, P. A., Schmidt, B. P., and Axelrod, T. S. 2002c, GRB Circular Network, 1219, 1

Prilutskii, O. F. and Usov, V. V. 1975, A&SS, 34, 395

Prochaska, J. X. *et al.* . 2004, ArXiv Astrophysics e-prints

Ramirez-Ruiz, E., Trentham, N., and Blain, A. W. 2002, MNRAS, 329, 465

Rauch, M. 1998, ARAA, 36, 267

Readhead, A. C. S. 1994, ApJ, 426, 51

Rees, M. J. and Meszaros, P. 1994, ApJ, 430, L93

Reeves, J. N. *et al.* . 2002, Nature, 416, 512

Reichart, D. E. 1999, ApJ, 521, L111

—. 2001, ApJ, 553, 235

Reichart, D. E. and Price, P. A. 2002, ApJ, 565, 174

Reichart, D. E. and Yost, S. A. 2001, ArXiv Astrophysics e-prints

Rhoads, J. E. 1997, ApJ, 487, L1+

—. 1999, ApJ, 525, 737

Rhoads, J. E. and Fruchter, A. S. 2001, ApJ, 546, 117

Richards, E. A. 2000, PASP, 112, 1001

Richards, E. A., Fomalont, E. B., Kellermann, K. I., Windhorst, R. A., Partridge, R. B., Cowie, L. L., and Barger, A. J. 1999, ApJ, 526, L73

Ricker, G. *et al.* . 2002, GRB Circular Network, 1220, 1

Rigon, L. *et al.* . 2003, MNRAS, 340, 191

Rola, C. S., Terlevich, E., and Terlevich, R. J. 1997, MNRAS, 289, 419

Rossi, E., Lazzati, D., and Rees, M. J. 2002, MNRAS, 332, 945

Ruderman, M. A., Tao, L., and Kluźniak, W. 2000, ApJ, 542, 243

Rybicki, G. B. and Lightman, A. P. 1979, Radiative processes in astrophysics (New York, Wiley-Interscience, 1979. 393 p.)

Sagar, R., Mohan, V., Pandey, S. B., Pandey, A. K., Stalin, C. S., and Castro Tirado, A. J. 2000, Bulletin of the Astronomical Society of India, 28, 499

Sahu, K. C. *et al.* . 1997, Nature, 387, 476

Sanders, D. B. and Mirabel, I. F. 1996, ARAA, 34, 749

Sari, R. 1997, ApJ, 489, L37+

Sari, R. and Esin, A. A. 2001, ApJ, 548, 787

Sari, R. and Mészáros, P. 2000, ApJ, 535, L33

Sari, R., Narayan, R., and Piran, T. 1996, ApJ, 473, 204

Sari, R., Piran, T., and Halpern, J. P. 1999, ApJ, 519, L17

Sari, R., Piran, T., and Narayan, R. 1998, ApJ, 497, L17+

Savaglio, S., Fall, S. M., and Fiore, F. 2003, ApJ, 585, 638

Schaefer, B. E. 1986, Advances in Space Research, 6, 47

Schaefer, B. E. *et al.* . 1989, ApJ, 340, 455

Schechter, P. L., Mateo, M., and Saha, A. 1993, PASP, 105, 1342

Schlegel, D. J., Finkbeiner, D. P., and Davis, M. 1998, ApJ, 500, 525

Schlegel, E. M. and Kirshner, R. P. 1989, AJ, 98, 577

Schmidt, M. 2001, ApJ, 552, 36

Schmidt, M., Higdon, J. C., and Hueter, G. 1988, ApJ, 329, L85

Scott, S. E. *et al.* . 2002, MNRAS, 331, 817

Sedov, L. I. 1946, Prikl. Mat. i Mekh., 10, 241

Shapley, A. E., Steidel, C. C., Pettini, M., and Adelberger, K. L. 2003, ApJ, 588, 65

Shemi, A. and Piran, T. 1990, ApJ, 365, L55

Sheth, K., Frail, D. A., White, S., Das, M., Bertoldi, F., Walter, F., Kulkarni, S. R., and Berger, E. 2003, Submitted to ApJ

Smail, I., Ivison, R. J., and Blain, A. W. 1997, ApJ, 490, L5+

Smail, I., Ivison, R. J., Blain, A. W., and Kneib, J.-P. 2002, MNRAS, 331, 495

Smartt, S. J., Vreeswijk, P. M., Ramirez-Ruiz, E., Gilmore, G. F., Meikle, W. P. S., Ferguson, A. M. N., and Knapen, J. H. 2002, ApJ, 572, L147

Smette, A., Fruchter, A., Gull, T., Sahu, K., Ferguson, H., Petro, L., and Lindler, D. 2000, GRB Circular Network, 603, 1

Smith, D. A. et al. . 2002a, ApJS, 141, 415

Smith, I. A. et al. . 1999, A&A, 347, 92

Smith, I. A., Tilanus, R. P. J., Wijers, R. A. M. J., Tanvir, N., Vreeswijk, P., Rol, E., and Kouveliotou, C. 2001, A&A, 380, 81

Smith, J. A. et al. . 2002b, AJ, 123, 2121

Soderberg, A. M., Kulkarni, S. R., Berger, E., Fox, D. B., and Sako, M. 2004, Submitted

Soderberg, A. M. and Ramirez-Ruiz, E. 2003, MNRAS, 345, 854

Soifer, B. T. et al. . 1984, ApJ, 283, L1

Sokolov, V. V. et al. . 2001, A&A, 372, 438

Sokolov, V. V., Kopylov, A. I., Zharikov, S. V., Feroci, M., Nicastro, L., and Palazzi, E. 1998, A&A, 334, 117

Sollerman, J., Kozma, C., Fransson, C., Leibundgut, B., Lundqvist, P., Ryde, F., and Woudt, P. 2000, ApJ, 537, L127

Spergel, D. N. et al. . 2003, ApJS, 148, 175

Sramek, R. A., Panagia, N., and Weiler, K. W. 1984, ApJ, 285, L59

Stanek, K. Z., Garnavich, P. M., Kaluzny, J., Pych, W., and Thompson, I. 1999, ApJ, 522, L39

Stanek, K. Z. et al. . 2003, ApJ, 591, L17

Stecker, F. W. and Frost, K. J. 1973, Nature Physical Science, 245, 70

Steidel, C. C., Adelberger, K. L., Giavalisco, M., Dickinson, M., and Pettini, M. 1999, ApJ, 519, 1

Steidel, C. C., Adelberger, K. L., Shapley, A. E., Pettini, M., Dickinson, M., and Giavalisco, M. 2003, ApJ, 592, 728

Steidel, C. C. and Sargent, W. L. W. 1992, ApJS, 80, 1

Steidel, C. C., Shapley, A. E., Pettini, M., Adelberger, K. L., Erb, D. K., Reddy, N. A., and Hunt, M. P. 2004, ApJ, 604, 534

Stern, B. E., Tikhomirova, Y., Kompaneets, D., Svensson, R., and Poutanen, J. 2001, ApJ, 563, 80

Storrie-Lombardi, L. J. and Wolfe, A. M. 2000, ApJ, 543, 552

Strong, I. B., Klebesadel, R. W., and Olson, R. A. 1974, ApJ, 188, L1+

Swartz, D. A. and Wheeler, J. C. 1991, ApJ, 379, L13

Swift, B., Li, W. D., and Filippenko, A. V. 2001, in International Astronomical Union Circular, 1–+

Takeshima, T., Markwardt, C., Marshall, F., Giblin, T., and Kippen, R. M. 1999, GRB Circular Network, 478, 1

Talbot, R. J. 1976, ApJ, 205, 535

Tan, J. C., Matzner, C. D., and McKee, C. F. 2001, ApJ, 551, 946

Taylor, G., Frail, D. A., Berger, E., and Kulkarni, S. R. 2004, submitted to ApJ

Taylor, G. B., Bloom, J. S., Frail, D. A., Kulkarni, S. R., Djorgovski, S. G., and Jacoby, B. A. 2000, ApJ, 537, L17

Taylor, G. I. 1950, Proc. R. Soc. London A, 201, 159

Taylor, J. H. and Cordes, J. M. 1993, ApJ, 411, 674

Taylor, J. R. 1982, An introduction to error analysis. The study of uncertainties in physical measurements (A Series of Books in Physics, Oxford: University Press, and Mill Valley: University Science Books, 1982)

Testa, V. *et al.* . 2003, GRB Circular Network, 2141, 1

Tiengo, A., Mereghetti, S., Ghisellini, G., Rossi, E., Ghirlanda, G., and Schartel, N. 2003, ArXiv Astrophysics e-prints, 5564

Tody, D. 1993, in ASP Conf. Ser. 52: Astronomical Data Analysis Software and Systems II, 173–+

Totani, T. 2003, ApJ, 598, 1151

Totani, T. and Panaitescu, A. 2002, ApJ, 576, 120

Uomoto, A. 1986, ApJ, 310, L35

Usov, V. V. 1992, Nature, 357, 472

Usov, V. V. and Chibisov, G. V. 1975, Soviet Astronomy, 19, 115

van den Bergh, S. 1983, A&SS, 97, 385

van Dyk, S. D., Sramek, R. A., Weiler, K. W., and Panagia, N. 1993, ApJ, 409, 162

van Paradijs, J. *et al.* . 1997, Nature, 386, 686

Vanderspek, R., Marshall, H. L., Ford, P. G., and Ricker, G. R. 2002, GRB Circular Network, 1504, 1

von Neumann, J. 1947, Los Alamos Sci. Lab. Tech. Ser., 7

Vrba, F. J. *et al.* . 2000, ApJ, 528, 254

Vreeswijk, P. M. and et al. 2004, astro-ph/0403080

Vreeswijk, P. M., Fender, R. P., Garrett, M. A., Tingay, S. J., Fruchter, A. S., and Kaper, L. 2001a, A&A, 380, L21

Vreeswijk, P. M. *et al.* . 2001b, ApJ, 546, 672

—. 1999, ApJ, 523, 171

Wade, R. A., Hoessel, J. G., Elias, J. H., and Huchra, J. P. 1979, PASP, 91, 35

Walker, M. A. 1998, MNRAS, 294, 307

Wang, L., Baade, D., Höflich, P., and Wheeler, J. C. 2003, ApJ, 592, 457

Wang, L., Howell, D. A., Höflich, P., and Wheeler, J. C. 2001, ApJ, 550, 1030

Watson, D., Reeves, J. N., Osborne, J. P., Tedds, J. A., O'Brien, P. T., Tomas, L., and Ehle, M. 2002, A&A, 395, L41

Waxman, E. 1997, ApJ, 489, L33+

—. 2004a, ApJ, 605, L97

—. 2004b, ApJ, 602, 886

Waxman, E. and Draine, B. T. 2000, ApJ, 537, 796

Waxman, E., Kulkarni, S. R., and Frail, D. A. 1998, ApJ, 497, 288

Webb, T. M. *et al.* . 2003, ApJ, 582, 6

Wei, D. M. and Lu, T. 2002, MNRAS, 332, 994

Weiler, K. W., Panagia, N., and Montes, M. J. 2001, ApJ, 562, 670

Weiler, K. W., Sramek, R. A., Panagia, N., van der Hulst, J. M., and Salvati, M. 1986, ApJ, 301, 790

Weiler, K. W., van Dyk, S. D., Montes, M. J., Panagia, N., and Sramek, R. A. 1998, ApJ, 500, 51

Wheaton, W. A. *et al.* . 1973, ApJ, 185, L57+

Wijers, R. A. M. J., Bloom, J. S., Bagla, J. S., and Natarajan, P. 1998, MNRAS, 294, L13

Wijers, R. A. M. J. and Galama, T. J. 1999, ApJ, 523, 177

Wilson, G., Cowie, L. L., Barger, A. J., and Burke, D. J. 2002, AJ, 124, 1258

Windhorst, R. A., Fomalont, E. B., Partridge, R. B., and Lowenthal, J. D. 1993, ApJ, 405, 498

Woosley, S. E. 1993, ApJ, 405, 273

Woosley, S. E., Eastman, R. G., and Schmidt, B. P. 1999, ApJ, 516, 788

Woosley, S. E., Langer, N., and Weaver, T. A. 1993, ApJ, 411, 823

Woosley, S. E. and Weaver, T. A. 1986, ARAA, 24, 205

Xu, D. W. and Qiu, Y. L. 2001, in International Astronomical Union Circular, 2–+

Yost, S. A. *et al.* . 2002, ApJ, 577, 155

Yost, S. A., Harrison, F. A., Sari, R., and Frail, D. A. 2003, ApJ, 597, 459

Young, T. R., Baron, E., and Branch, D. 1995, ApJ, 449, L51+

Yun, M. S. and Carilli, C. L. 2002, ApJ, 568, 88

Zhang, B. and Mészáros, P. 2002, ApJ, 571, 876

Zwicky, F. 1958, PASP, 70, 506

—. 1974, A&SS, 28, 111